BURSIAN

Environmental Health Criteria 58

SELENIUM

Published under the joint sponsorship of
the United Nations Environment Programme,
the International Labour Organisation,
and the World Health Organization

World Health Organization
Geneva, 1987

The International Programme on Chemical Safety (IPCS) is a joint venture of the United Nations Environment Programme, the International Labour Organisation, and the World Health Organization. The main objective of the IPCS is to carry out and disseminate evaluations of the effects of chemicals on human health and the quality of the environment. Supporting activities include the development of epidemiological, experimental laboratory, and risk-assessment methods that could produce internationally comparable results, and the development of manpower in the field of toxicology. Other activities carried out by IPCS include the development of know-how for coping with chemical accidents, coordination of laboratory testing and epidemiological studies, and promotion of research on the mechanisms of the biological action of chemicals.

ISBN 92 4 154258 6

PRINTED IN FINLAND
85/6657 — VAMMALA — 5800

CONTENTS

Page

ENVIRONMENTAL HEATH CRITERIA FOR SELENIUM

1. SUMMARY AND RECOMMENDATIONS FOR FURTHER RESEARCH ... 13

   1.1 Summary 13
       1.1.1 Properties and analytical methods 13
       1.1.2 Sources, environmental transport, and
              distribution 13
       1.1.3 Environmental levels and exposures 14
       1.1.4 Selenium metabolism 15
            1.1.4.1 Absorption 15
            1.1.4.2 Total human body selenium
                    content 16
            1.1.4.3 Distribution 16
            1.1.4.4 Metabolic pools of selenium in
                    the body 16
            1.1.4.5 Metabolic conversion 17
            1.1.4.6 Effect of chemical form of
                    selenium on its metabolism 17
            1.1.4.7 Selenium excretion 18
       1.1.5 Effects on animals 18
            1.1.5.1 Selenium toxicity 18
            1.1.5.2 Selenium deficiency 20
       1.1.6 Effects on man 22
            1.1.6.1 General population exposure ... 22
            1.1.6.2 Occupational exposure 23
   1.2 Recommendations for further activities 23

2. CHEMICAL AND PHYSICAL PROPERTIES; ANALYTICAL METHODS . 25

   2.1 Properties 25
   2.2 Analytical methods 27
       2.2.1 Sample collection, processing, and storage 27
       2.2.2 Sample decomposition or other preliminary
               treatment 28
            2.2.2.1 Wet digestion 28
            2.2.2.2 Predigestion 29
            2.2.2.3 Combustion 29
            2.2.2.4 Fusion 29
            2.2.2.5 Concentration 30
       2.2.3 Removal from interfering substances 30
       2.2.4 Detection and identification of selenium . 30
       2.2.5 Measurement of selenium 30
            2.2.5.1 Fluorometric analysis 31

Page

2.2.5.2 Neutron activation analysis . . . 32
2.2.5.3 Atomic absorption spectrometry . . 33
2.2.5.4 Other methods 34

3. SOURCES, TRANSPORT, AND CYCLING OF SELENIUM IN THE
   ENVIRONMENT 36

   3.1 Natural sources 36
        3.1.1 Rocks and soils 36
        3.1.2 Natural selenium in the food chain 37
        3.1.3 Water and air 38
   3.2 Man-made sources 38
        3.2.1 Agriculture 38
        3.2.2 Industry 39
   3.3 Environmental transport 41
   3.4 Biological selenium cycle 41

4. LEVELS IN ENVIRONMENTAL MEDIA 44

   4.1 Levels and chemical forms of selenium in food . . 44
        4.1.1 Levels in food 44
               4.1.1.1 Natural differences among food
                        commodities 44
               4.1.1.2 Effects of natural differences in
                        the availability of selenium in the
                        environment on levels in food . . 47
               4.1.1.3 Man-induced changes in selenium
                        levels in food 50
        4.1.2 Chemical forms of selenium in food 53
   4.2 Drinking-water 54
   4.3 Air 55

5. HUMAN EXPOSURE 56

   5.1 Estimate of general population exposure 56
        5.1.1 Food 56
               5.1.1.1 Geographical variation 56
               5.1.1.2 Food habits (consumption
                        patterns) 60
               5.1.1.3 Elderly people 61
               5.1.1.4 Infants and children 61
               5.1.1.5 Special medical diets 63
   5.2 Occupational exposure 63
        5.2.1 Levels in the work-place air 63
        5.2.2 Biological monitoring 65

6. METABOLISM OF SELENIUM 67

Page

6.1 Absorption 67
    6.1.1 Gastrointestinal absorption 67
          6.1.1.1 Animal studies 67
          6.1.1.2 Human studies 68
    6.1.2 Absorption by inhalation 69
    6.1.3 Absorption through the skin 70
6.2 Distribution in the organism 71
    6.2.1 Transport 71
    6.2.2 Organs 71
          6.2.2.1 Animal studies 71
          6.2.2.2 Human studies 73
    6.2.3 Blood 73
    6.2.4 Total-body selenium content 77
6.3 Excretion in urine, faeces, and expired air . . . 78
    6.3.1 Animal studies 78
    6.3.2 Human studies 80
          6.3.2.1 Excretion of selenium 80
          6.3.2.2 Balance studies 81
6.4 Retention and turnover 82
    6.4.1 Animal studies 82
    6.4.2 Controlled human studies 86
6.5 Metabolic transformation 87
    6.5.1 Animal studies 88
          6.5.1.1 Reduction and methylation 88
          6.5.1.2 Form in proteins 89
          6.5.1.3 Conversion of selenium compounds
                  to nutritionally-active forms of
                  selenium 89
    6.5.2 Human studies 90

7. EFFECTS OF SELENIUM ON ANIMALS 91

7.1 Selenium toxicity 91
    7.1.1 Farm animal diseases associated with a high
          selenium intake 91
    7.1.2 Toxicity in experimental animals 92
          7.1.2.1 Acute and subacute toxicity -
                  single or repeated exposure studies
                  with oral, intraperitoneal, or
                  cutaneous administration 93
          7.1.2.2 Effects of long-term oral
                  exposure 96
          7.1.2.3 Inhalation toxicity 103
    7.1.3 Blood levels in toxicity 105
    7.1.4 Effects on reproduction 105
    7.1.5 Effects on dental caries 111
    7.1.6 Factors influencing toxicity 114

Page

7.1.6.1 Form of selenium 114
7.1.6.2 Nutritional factors 114
7.1.6.3 Arsenic 115
7.1.6.4 Sulfate 115
7.1.6.5 Adaptation 115
7.1.7 Mechanism of toxicity 117
7.2 Selenium deficiency 118
7.2.1 Animal diseases 118
7.2.2 Intakes needed to prevent deficiency . . . 119
7.2.2.1 Quantitative dietary levels . . . 119
7.2.2.2 Bioavailability 120
7.2.3 Blood and tissue levels in deficiency . . . 120
7.2.4 Physiological role: glutathione
peroxidase 121
7.2.4.1 Function of selenium and
relationship to vitamin E 121
7.2.4.2 Effect of selenium intake on
tissue-glutathione peroxidase
activity 122
7.2.4.3 Relationship between blood-
selenium levels and erythrocyte-
glutathione peroxidase activity . 124
7.2.4.4 Gluthathione peroxidase as an
indicator of selenium status . . . 125
7.2.5 Other possible physiological roles 127
7.2.5.1 Homeostasis of hepatic haem . . . 127
7.2.5.2 Microsomal and mitochondrial
electron transport 128
7.2.5.3 The immune response 129
7.2.5.4 Selenium and vision 130
7.2.6 Effects on reproduction 131
7.2.7 Factors influencing deficiency 135
7.2.7.1 Form of selenium 135
7.2.7.2 Vitamin E 136
7.2.7.3 Heavy metals and other minerals . 137
7.2.7.4 Xenobiotics 138
7.2.7.5 Exercise stress 141
7.3 Ratio between toxic and sufficient exposures . . . 142
7.4 Protection against heavy metal toxicity 143
7.4.1 Mercury 143
7.4.2 Cadmium 143
7.4.3 Other heavy metals 144
7.5 Cytotoxicity, mutagenicity, and anti-
mutagenicity 144
7.5.1 Cytotoxicity and mutagenicity 144
7.5.2 Anti-mutagenicity 149
7.6 Teratogenicity 154

Page

7.7 Carcinogenicity and anti-carcinogenicity 155
     7.7.1 Selenium as a possible carcinogen 155
     7.7.2 Selenium as a possible anti-carcinogen . . 164

8. EFFECTS OF SELENIUM ON MAN 175

8.1 High selenium intake 175
     8.1.1 General population 175
            8.1.1.1 Signs and symptoms 175
            8.1.1.2 Attempts to associate high
                     selenium intake with human
                     diseases 183
     8.1.2 Reports on health effects associated with
            occupational exposure 185
            8.1.2.1 Fumes and dust of selenium and
                     its compounds 186
                     8.1.2.1.1 Selenium dioxide . . . 187
                     8.1.2.1.2 Hydrogen selenide . . 194
                     8.1.2.1.3 Selenium oxychloride . . 198
8.2 Low selenium intake 198
     8.2.1 Evidence supporting the possible essential-
            ity of selenium in man 198
     8.2.2 Signs and symptoms of low intake 199
     8.2.3 Dietary levels consistent with good
            nutrition 201
            8.2.3.1 Quantitative estimates 201
            8.2.3.2 Nutritional bioavailability . . 203
     8.2.4 Blood and urine levels typical of low
            intake 205
     8.2.5 Relationship between blood-selenium levels
            and erythrocyte-glutathione peroxidase
            activity 206
     8.2.6 Attempts to associate low selenium intake
            with human diseases 209
            8.2.6.1 Keshan disease 209
            8.2.6.2 Kashin-Beck disease 213
            8.2.6.3 Cancer 214
            8.2.6.4 Heart disease 221

9. EVALUATION OF THE HEALTH RISKS ASSOCIATED WITH EXCESSIVE
    OR DEFICIENT SELENIUM EXPOSURE 224

9.1 The need to consider the essentiality of selenium
     in health risk evaluation 224
9.2 Pathway of selenium exposure for the general
     population 224

9.3 Quantitative assessment of human selenium
     exposure 225
     9.3.1 Analytical methods for selenium 225
     9.3.2 Food intake data 225
     9.3.3 Blood-selenium 226
     9.3.4 Hair-selenium 226
     9.3.5 Urine-selenium 227
     9.3.6 Blood-glutathione peroxidase 227
9.4 Levels of dietary selenium exposure in the general
     population 228
9.5 Evaluation of health risks - general population . 229
     9.5.1 Predictive value of animal studies 229
     9.5.2 Studies on high-exposure effects in the
            general population 231
     9.5.3 Studies on low-exposure effects in the
            general population 232
     9.5.4 Evaluation of the involvement of selenium
            in human diseases of multiple
            etiopathogenesis 233
            9.5.4.1 Keshan disease 233
            9.5.4.2 Kashin-Beck disease 233
            9.5.4.3 Ischaemic heart disease 234
            9.5.4.4 Studies on the involvement of
                     selenium in cancer 234
            9.5.4.5 Caries 236
            9.5.4.6 Health effects related to
                     reproduction 236
9.6 Occupational exposure 236

REFERENCES 238

WHO TASK GROUP ON SELENIUM

Members

Dr R.F. Burk, Jr, Department of Medicine, University of Texas
Health Science Center, San Antonio, Texas, USA

Professor A.I. Diplock, Department of Biochemistry, Guy's
Hospital Medical School, London, United Kingdom (Chairman)

Dr H.N.B. Gopalan, University of Nairobi, Department of
Botany, Nairobi, Kenya

Dr J. Chen, Department of Nutrition and Food Hygiene,
Institute of Health, China National Centre for Preventive
Medicine, Beijing, People's Republic of China

Dr G.N. Krasovskij, Sysir Institute of General and Community
Hygiene, Academy of Medical Sciences of the USSR, Moscow,
USSR[a]

Professor C.R. Krishna Murti, Integrated Environmental
Programme on Heavy Metals, Centre for Environmental
Studies, Anna University, College of Engineering, Guindy,
Madras, India[a]

Dr O.A. Levander, Vitamin and Mineral Nutrition Institute, US
Department of Agriculture, Beltsville, Maryland, USA
(Rapporteur)

Professor A. Massoud, Department of Community, Environmental
and Occupational Medicine, Faculty of Medicine, Ain Shams
University, Cairo, Egypt[a]

Professor M.F. Robinson, Nutrition Department, University of
Otago, Dunedin, New Zealand

Secretariat

Dr J. Parizek, International Programme on Chemical Safety,
World Health Organization, Geneva, Switzerland (Secretary)

Dr E.S. Johnson, International Agency for Research on Cancer,
Lyons, France

---

[a] Invited but unable to attend.

NOTE TO READERS OF THE CRITERIA DOCUMENTS

Every effort has been made to present information in the criteria documents as accurately as possible without unduly delaying their publication. In the interest of all users of the environmental health criteria documents, readers are kindly requested to communicate any errors that may have occurred to the Manager of the International Programme on Chemical Safety, World Health Organization, Geneva, Switzerland, in order that they may be included in corrigenda, which will appear in subsequent volumes.

*   *   *

A detailed data profile and a legal file can be obtained from the International Register of Potentially Toxic Chemicals, Palais des Nations, 1211 Geneva 10, Switzerland (Telephone no. 988400 - 985850).

ENVIRONMENTAL HEALTH CRITERIA FOR SELENIUM

A WHO Task Group on Environmental Health Criteria for Selenium was held in Geneva on 2 - 6 December 1985. Dr J. Parizek opened the meeting on behalf of the Director-General. The Task Group reviewed and revised the draft criteria document and made an evaluation of the health risks of exposure to selenium.
DR O.A. LEVANDER of the US Department of Agriculture was responsible for the preparation of the drafts of the document. The efforts of all who helped in the preparation and finalization of the document are gratefully acknowledged.

* * *

Partial financial support for the publication of this criteria document was kindly provided by the United States Department of Health and Human Services, through a contract from the National Institute of Environmental Health Sciences, Research Triangle Park, North Carolina, USA - a WHO Collaborating Centre for Environmental Health Effects.

1. SUMMARY AND RECOMMENDATIONS FOR FURTHER RESEARCH

1.1 Summary

1.1.1 Properties and analytical methods

Selenium exists naturally in several oxidation states and some of its chemical forms are volatile. Many selenium analogues of organic sulfur compounds exist in nature. A number of procedures exist for the determination of selenium. However, as selenium exists in volatile and unstable forms and because of the inhomogeniety of many types of materials, the methods of sampling, preparation, and storage are equally as important as the analytical methods. Great care is necessary to avoid contamination or loss of the element.

The most commonly used analytical methods depend on wet digestion for destroying organic matter and freeing the element. Some procedures depend on the formation of the piazoselenol; this is extracted in an organic solvent and the fluorescence determined. The fluorometric method with a wide variety of modifications is very sensitive and reliable and can be adapted for most materials. It is also inexpensive. Atomic absorption spectrometry, especially with the atomization of the selenium as the hydride, has been useful, and atomic absorption methods based on the Zeeman effect background correction show promise for the determination of selenium in biological matrices, without prior sample digestion. Neutron activation analysis, particularly when combined with the chemical separation of selenium, is an excellent method, limited only by the cost and availability of equipment.

1.1.2 Sources, environmental transport, and distribution

Selenium appears to be ubiquitous. However, its uneven distribution over the face of the earth results in regions with very low or very high natural levels of selenium in the environment. Geophysical, biological, and industrial processes are involved in the distribution and transport of the element and its cycling, but the relative importance of these processes has not been established. However, the natural geophysical and biological processes are probably almost entirely responsible for the present status of selenium in the general environment. This must be given primary consideration in any evaluation of the superimposed effects of man's activity on selenium in the environment and food chains. Some human activities are responsible for the redistribution of selenium in the environment. Industrial sources of selenium stem initially from copper refining. During this refining and the purification of the selenium, there can be some loss of the element into the environment. In addition, industries concerned with the production of glass, electronic equipment, or certain metals may emit selenium into the environment in the immediate vicinity of the factories involved. The inclusion of the element in manufactured products provides another avenue for its distribution. There is concern in several countries with regard to the possible health effects of low and/or decreasing levels of selenium in the soil (sections 1.1.3, 1.1.5.2 and 1.1.6.1); the use of fertilizers containing selenium compounds in some Nordic countries is a remarkable example of intentional human intervention in the environmental distribution of selenium.

1.1.3   Environmental levels and exposures

The range of selenium levels in different foods can vary widely, depending on the natural availability of selenium in the environment and on certain activities of man, such as the direct addition of selenium to the food supply.

No data are available on the chemical forms of selenium in foods produced under normal conditions.

The limited analytical data available show that the levels of selenium typically found in foods (on a wet-weight basis) range from: 0.4 - 1.5 mg/kg in liver, kidney, and seafood; 0.1 - 0.4 mg/kg in muscle meat; < 0.1 - 0.8 mg/kg or more in cereals or cereal products; < 0.1 - 0.3 mg/kg in dairy products; and < 0.1 mg/kg in most fruits and vegetables. However, in countries with low selenium levels in soil, lower selenium levels than the above were reported, in particular in meats, cereals, and dairy products. The levels of selenium in baby foods tend to follow the same trends as in adult foods, i.e., meat and cereal products contain the highest levels, and fruit and vegetable products the lowest. Meat-based infant formulae contain more selenium than formulae based on milk or soy protein. Average selenium concentrations in human milk range from 0.013 to 0.018 mg/litre. However, in countries with low selenium levels in soil and food, lower selenium levels in breast milk were reported.

Special medical diets based on egg albumen contain more selenium than diets based on casein hydrolysate. Chemically-defined diets or total parenteral nutrition solutions based on amino acid mixtures contain very low levels of selenium.

Levels of selenium in air and water are usually very low, i.e., less than 10 ng/m^3 in air and only a few µg/litre water.

Food constitutes the main route of exposure to selenium for the general population. Because of geochemical differences, the estimates of adult human exposure to selenium via the diet range from 11 to 5000 μg/day, in different parts of the world; however, dietary intake more usually falls within the range of 20 - 300 μg/day. Food consumption patterns also affect dietary-selenium intake, and the extreme values observed have occurred in populations consuming a monotonous diet comprising a limited range of locally-grown staple foods. The estimated selenium intake of infants in different parts of the world during the first month of life ranges from 5 to 55 μg/day, because of the variation in levels of selenium in milk. Children consuming synthetic diets, as part of long-term diet therapy for certain metabolic diseases, such as phenylketonuria and maple syrup urine disease, ingest only 5 - 11 μg/day. Adults, maintained on chemically-defined diets or total parenteral nutrition solutions based on amino acid mixtures, receive only 1 or 2 μg/day. There is growing evidence of the importance of the bioavailability of selenium in different foods for human beings.

Human occupational exposure to selenium is primarily via the air. Exposure via direct contact is rarely of importance, unless local irritation or skin damage caused by vesicant selenium compounds facilitates cutaneous absorption. Few quantitative data are available on the actual levels of occupational exposure to selenium. However, it is likely that analyses carried out several decades ago yielded higher values (several mg/m³) than those carried out more recently.

1.1.4   Selenium metabolism

1.1.4.1 Absorption

Potential sites of selenium absorption from the environment are the gastrointestinal tract, the respiratory tract, and the skin. Limited animal and no quantitative human data are available on the pulmonary absorption of selenium. Nevertheless, high urinary-selenium levels in workers exposed to selenium in air have been reported and could indicate pulmonary absorption. Selenite and selenium oxychloride have been shown to be absorbed through the skin of experimental animals. The assessment of the possible pulmonary or dermal absorption rates for various selenium compounds must take into account differences in their physical and chemical properties.

The results of several experimental animal studies as well as direct measurements of selenium absorption in man indicate that selenium compounds can be readily absorbed in the intestinal tract and there appears to be no physiological control over this absorption. This statement is based largely on studies of ^{75}Se-labelled selenite absorption in rats fed widely varying amounts of selenium in the diet. All groups absorbed more than 95% of the ^{75}Se administered by stomach tube. High absorption of selenium was reported in women given 1-mg doses of selenium as selenomethionine or selenite dissolved in water.

1.1.4.2 Total human body selenium content[a]

The reported estimates of the amount of selenium in the adult human body range from 3 to 14.6 mg. The lower estimates, 3.0 or 6.1 mg, were obtained in New Zealand by indirect techniques following the administration of radiolabelled selenite or selenomethionine, respectively, to healthy human volunteers. The higher estimate, 14.6 mg, was obtained in the USA and was based on direct analysis of autopsy material accounting for 91.7% of the body mass. The difference between these 2 estimates may be due either to differences in the techniques employed or to differences in dietary-selenium intake in New Zealand and the USA.

1.1.4.3 Distribution

Under normal conditions, levels of selenium are higher in the kidney and liver than in the other major body tissues. Although muscle-selenium levels are lower, muscle is the tissue present in the greatest amount in the body, and thus accounts for the highest proportion of the total body selenium. It is generally assumed that the levels of selenium in the above tissues and in red blood cells and plasma are related to the total body content of selenium, but additional studies are needed to establish this point under various conditions of selenium exposure, and also in man.

1.1.4.4 Metabolic pools of selenium in the body

There is no evidence of a regulated or specific storage form of selenium, analogous to ferritin-bound iron. However, various data suggest that several body pools of selenium could be considered as serving a similar purpose. Evidence for a carry-over effect of selenium exists. Sheep fed plants containing adequate levels of selenium for 5 months and then fed low-selenium plants for 10 months produced lambs with higher tissue-selenium contents and fewer signs of selenium deficiency than sheep fed only low-selenium plants. North

---

[a] The Task Group felt that the term "body burden" might be misleading when applied to an essential trace element.

Americans moving to the low-selenium area of New Zealand experienced a slow drop in blood-selenium contents for approximately 1 year, before reaching New Zealand levels. These observations could be explained by sequestration of selenium in the form of selenomethionine and/or other selenoamino acids incorporated into the primary structure of proteins throughout the body. Selenium sequestered in this form would be released and made available at a rate corresponding to the turnover of proteins and the catabolism of selenomethionine. This mechanism could provide selenium for animals temporarily unable to obtain it in sufficient amounts through their diets.

The Task Group concluded that present knowledge about selenium biochemistry was consistent with the concept of different selenium pools in the body, even if direct measurements of the quantitative aspects regarding the compartmentalization of selenium in the human body and the related turnover rates were not available.

1.1.4.5 Metabolic conversion

Biological utilization of selenium from different chemical forms has usually been assessed by determining the ability of such compounds to prevent selenium-deficiency diseases or to increase glutathione peroxidase activity in selenium-deficient animals. Selenite and selenomethionine readily fulfil both these criteria and can thus be considered convertible to meta-bolically-active forms. In addition, it should be recognized that, in several of the early metabolic studies on selenium, many other selenium-containing compounds were shown to prevent dietary liver necrosis and thus to undergo metabolic conversion to nutritionally-active forms.

Some of the excretory metabolites of selenium have been identified. Dimethylselenide is excreted in the breath at high levels of selenium exposure. Trimethylselenonium ion and several unidentified selenium metabolites are excreted in the urine.

1.1.4.6 Effect of chemical form of selenium on its metabolism

Animal studies indicate that selenite is not converted to selenomethionine in the body. Nevertheless, both selenite and selenomethionine are able to satisfy the nutritional requirement for selenium.

The results of several studies indicate differences in the metabolism of these two forms of selenium. On the basis of limited human studies, in which 100 µg selenium/day was given, selenium in the form of selenomethionine was retained in larger amounts than selenium given as selenite; it also

2

resulted in higher blood-selenium levels. Comparison of selenite and selenomethionine in chick studies demonstrated that retention of selenium was greater when given as seleno-methionine than when given as selenite. When mice were given toxic levels of selenium compounds, selenomethionine admini-stration resulted in higher tissue-selenium levels than administration of the same level of selenium as selenite. Furthermore, manifestations of toxicity disappeared rapidly when selenite was withdrawn from the diet but subsided only slowly following withdrawal of selenomethionine. The results of limited studies suggest that the metabolism of both selenium-methylselenocysteine and selenocystine resembles that of selenite rather than that of selenomethionine. These findings can be explained by the incorporation of a portion of the selenium supplied as selenomethionine into the tissue proteins as the amino acid.

1.1.4.7 Selenium excretion

In human volunteers given tracer doses of inorganic or organic selenium compounds orally, excretion was mainly by the urinary route. However, when people consumed naturally-occurring selenium in foods, approximately equal proportions were excreted in the urine and faeces. Very little was excreted in the sweat, and animal studies have shown that significant respiratory excretion of volatile selenium compounds only occurs in cases of very high selenium exposure. Selenium loss from the body, as judged by whole-body counting studies following administration of single doses of ^{75}Se-labelled compounds, consisted of an initial phase of rapid decrease of whole body radioactivity followed, after several days, by a phase of more gradual ^{75}Se excretion. The results of animal studies indicate that the rate of selenium excretion in the initial phase is affected by the dose of selenium administered as well as by selenium status, whereas the second phase of excretion is mainly affected by selenium status.

1.1.5 Effects on animals

1.1.5.1 Selenium toxicity

A characteristic sign of acute selenium poisoning in animals is the odour due to the pulmonary excretion of dimethyl selenide. Other signs of acute selenium poisoning in dogs and rats include: vomiting, dyspnoea, tetanic spasms, and death from respiratory failure. Pathological changes include congestion of the liver with areas of focal necrosis, congestion of the kidney, endocarditis, myocarditis, and peticheal haemorrhages of the epicardium. The oral LD_{50} reported for sodium selenite varies from 2.3 to 13 mg selenium/kg body weight because of species differences and other variables. Inhalation exposure to various selenium compounds, including selenium dioxide, hydrogen selenide, and selenium dust, proved to be toxic, causing damage to the respiratory tract, liver, and other organs, lethality being dependent on the level and duration of exposure. Because of the limited number of studies and differences in species and other experimental conditions used, evaluation of the dose-response relationships for the different compounds is difficult.

It has been shown that the amount of dietary selenium needed to cause chronic toxicity in animals is influenced by many variables including the form of selenium and the type of diet. When fed in the diet, elemental selenium has a low order of toxicity because of its insolubility. Sodium selenite or seleniferous wheat fed at a level of 6.4 mg selenium/kg diet caused growth inhibition, liver cirrhosis, and splenomegaly in young growing rats fed the diet for 6 weeks. Growth inhibition at 4.8 mg/kg diet was not statistically significant. Levels of 8 mg/kg diet or more caused additional effects such as pancreatic enlargement, anaemia, elevated serum-bilirubin levels and, in 4 weeks, death. In another study when rats were fed 16 mg selenium/kg diet as sodium selenate in a commercial "laboratory chow" type ration, their median survival age was reported to be 96 days, and the predominant histopathological lesion was acute toxic hepatitis. The predominant lesion at 8 mg/kg in the chow diet was chronic toxic hepatitis and the median survival age was 429 days. Only 4 mg/kg were needed to cause acute toxic hepatitis in rats consuming a semi-purified diet containing 12% casein. Increased activities of serum alkaline phosphatase (EC 3.1.3.1) and glutamic-pyruvic transaminase (EC 2.6.1.2) were observed in young growing rats fed a semi-purified diet containing 4.5 mg selenium/kg as seleniferous sesame meal.

By using other criteria of toxicity, some research workers have claimed deleterious effects of selenium at lower levels of intake, however, there is a need to develop and validate more sensitive and specific indicators of selenium poisoning. Farm animals raised in regions where there are high levels of available selenium in the soil develop toxicity diseases as a result of consuming plants containing excess selenium. The levels of dietary selenium needed to cause chronic toxicity are 5 mg/kg or more in cattle and 2 mg/kg or more in sheep. Blood-selenium levels higher than 2 mg/litre are generally associated with frank chronic selenium poisoning in cattle,

between 1 and 2 mg/litre. Blood-selenium levels of 0.6 -
0.7 mg/litre are associated with chronic selenosis in sheep.

On the basis of anecdotal reports, it has been suggested
that excess selenium may cause various practical reproduction
problems in farm animals, such as decreased reproductive performance in livestock or decreased hatchability in chickens
because of deformities, at levels that do not cause obvious
manifestations of toxicity. A marked deterioration in reproductive performance was noted in a multigeneration study in
which mice were given sodium selenate in the drinking-water at
3 mg selenium/litre, a level that had no effect on growth or
survival in a typical single-generation toxicity study. There
was a considerable decrease in the number of litters produced
by the third-generation mice and a considerable increase in
the number of runts among the mice that were born.

When monkeys were fed a cariogenic diet and given 2 mg
selenium/litre as sodium selenite for 15 months followed by
1 mg/litre for 45 months, the incidence of caries increased
when the selenium was given during tooth development but not
when the teeth were exposed post-eruptively. On the other
hand, the effect of a cariogenic diet given for two months
after weaning was decreased in rats whose mothers had been
given, during the second half of pregnancy and during
lactation, drinking water containing selenium at the level of
0.8 mg/litre in the form of sodium selenite or selenomethionine.

In two independent studies, it was shown that, under
certain circumstances, selenium compounds were shown to be
less toxic to animals kept on a high dietary intake of
selenium, thereby suggesting possible adaptation to high
selenium exposures.

1.1.5.2 Selenium deficiency

Animals raised in regions where there are low levels of
available selenium in the soil develop deficiency diseases
through consuming plants lacking adequate selenium. These
diseases can be prevented by feeding inorganic selenium
compounds to the animals.

Deficiency diseases in which both selenium and vitamin E
may play a role include nutritional muscular dystrophy in
sheep and cattle, exudative diathesis in chickens, and liver
necrosis in swine and rats. Signs specific for selenium
deficiency in the absence of vitamin E deficiency include
pancreatic degeneration in chicks, and poor growth, reproductive failure, vascular changes, and cataracts in rats.

The level of dietary selenium needed to prevent deficiency
diseases in animals depends on the vitamin E content of the
diet. For example, chicks receiving a diet deficient in

vitamin E need 0.05 mg selenium/kg diet to prevent exudative diathesis whereas 0.01 mg/kg will be sufficient to prevent the disease if the diet contains 100 mg vitamin E/kg. Under normal vitamin E intake, the level of dietary selenium needed to prevent deficiency is about 0.02 mg/kg for ruminants and 0.03 - 0.05 mg/kg for poultry. Blood-selenium levels of less than 0.05 mg/litre are usually associated with signs of selenium deficiency in sheep. Hepatic selenium levels of less than 0.21 mg/kg (dry basis) are associated with a high incidence of white muscle disease in lambs. However, levels indicative of marginal deficiency can be influenced by the vitamin E status of the animal.

The physiological function of selenium and its nutritional relationship to the biological antioxidant vitamin E can be largely explained on the basis of its role as a component of the enzyme glutathione peroxidase (EC 1.11.1.9), which is responsible for the destruction of hydrogen peroxide and lipid peroxides.

There is a close association between the level of dietary intake of selenium and the glutathione peroxidase activity in several organs. Also, the glutathione peroxidase activity of erythrocytes is closely associated with the selenium content of the blood. Because of these close relationships, measurement of glutathione peroxidase offers a convenient method for assessing selenium intake. However, the activity of the enzyme is influenced by several nutritional and environmental factors that must be considered when using it as an index of selenium status.

The ratio of the toxic level of selenium in the diet to the nutritional level of selenium in the diet is approximately 100, but this ratio can be decreased by nutritional or environmental factors. For example, deficiency of vitamin E increases the susceptibility of animals to selenium toxicity but also increases the nutritional need for the element. Furthermore, inorganic mercury potentiates the toxicity of methylated selenium metabolites, whereas methyl-mercury potentiates selenium deficiency.

Dietary selenium can protect against the toxicity of several heavy metals, such as mercury or cadmium, and certain xenobiotics, such as paraquat, but the mechanism of these protective effects is not known.

Selenium has been suspected of being a carcinogen in the past, but more recent research suggests that it may be able to protect against certain types of cancers in experimental animals.

1.1.6   Effects on man

1.1.6.1   General population exposure

As stated above, the main environmental pathway of selenium exposure in the general population is through food. Nutritional surveys have shown that extreme dietary intakes range from 11 - 5000 µg/day, but on most diets intakes between 20 and 300 µg/day can be considered as more typical. The extremes in intake are reflected in extreme levels of selenium in blood, mean reported values ranging from 0.021 to 3.2 mg/litre. The highest blood-selenium levels ever observed in the general population were found in an area of the People's Republic of China in which an episode of intoxication reported as selenosis had occurred some years earlier. In this respect, as well as in at least two other studies in over-exposed populations, hair loss and nail pathology were the most marked and readily documented toxic signs. The Task Group, being aware of the hepatotoxicity of selenium compounds observed in animal studies, noted that no clinical signs of hepatotoxicity were observed in the studies of people exposed to high levels of dietary selenium, but concluded that there is a need for more thorough evaluation of hepatic function in persons with high selenium exposure. Tooth decay was also observed in several studies on over-exposed populations, but in evaluating its significance the Task Group was unable to exclude interference by other environmental factors. The Task Group recognized the difficulty in establishing an exact dose-response with respect to selenium in the above studies. The range of blood values noted was 0.44 - 3.2 mg/litre; in these studies no adverse effects were reported at the lower level, whereas clear effects on the hair and nails were observed at and above a level of 0.813 mg/litre.

The lowest blood-selenium levels ever reported in a general population were seen in regions of the People's Republic of China where Keshan disease and Kashin-Beck disease are known to be endemic. The Task Group recognized the intensive research being carried out concerning the involvement of selenium in the multifactorial etiology of these diseases and the use of selenium compounds in their prevention.

The Task Group considered a large number of studies on the possible relationship between low levels of selenium intake and a high incidence of cancer. In evaluating the available data, the Task Group noted both consistencies and inconsistencies, which made it difficult to draw a firm conclusion. However, some recent case-control studies within prospective studies suggested the importance of this approach for a firmer evaluation of the involvement of the level of selenium intake in cancer prevention. The Task Group was aware of an intervention trial being carried out in China which could provide additional information on the association between selenium exposure and human cancer risk.

1.1.6.2 Occupational exposure

As discussed above, the main environmental pathway of occupational exposure to selenium is through the air, or in some cases, by direct dermal contact. The selenium compounds likely to be encountered include selenium dust, selenium dioxide, and hydrogen selenide. The toxicological potential of selenium for human beings employed in industry, can be inferred from respiratory exposure studies carried out on laboratory animals and from ad hoc case studies of industrial accidents.

Caution must be exercised when extrapolating the results of animal studies or industrial accidents to the industrial health aspects of long-term selenium exposure, because data available from animal studies dealing with respiratory exposure are limited, the exposure periods are brief, and industrial accidents involve situations in which the exposure was brief and at an undetermined level.

Extrapolation from general-population exposure is also quite difficult, since selenium ingested in the diet may have characteristics quite distinct from selenium encountered under occupational exposure conditions: the chemical and physical forms are likely to be different and these forms may change after contact with the moist mucous membranes or with sweat. Knowledge about the health effects of industrial selenium exposure is rudimentary, since acute exposures are accidental and must be described on an ad hoc case study basis. The Task Group did not know of any epidemiological investigations of the effects of long-term industrial exposure to selenium that included unexposed control groups, nor were long-term follow-up studies with appropriate control groups available. Exposure levels in short- and long-term exposure studies are often ill-defined and the form of selenium is not characterized. Confounding factors such as simultaneous exposure to other toxic materials may exist.

1.2 Recommendations for Further Activities

1. The Task Group recommended that further studies were needed on the well-identified population segments over-exposed to selenium to clarify the signs and symptoms of selenium overexposure in man and establish the relevant dose-response relationships. The exclusion of hepatotoxic effects seems to be of particular importance.

2. Further research on Keshan disease and Kashin-Beck disease should be undertaken, not only to improve prevention of these diseases, but also to provide basic information on the effects of low selenium intake in man.

3. Further research is needed on the relationship between the level of selenium intake and the incidence of cancer.

2. CHEMICAL AND PHYSICAL PROPERTIES; ANALYTICAL METHODS

2.1 Properties

Selenium belongs to group VI of the periodic table, between sulfur and tellurium. It has similar chemical and physical properties because the structure of its outer electron shells is similar. Some chemical and physical properties of selenium are listed in Table 1.

Table 1. Some chemical and physical properties of selenium[a]

Properties	Values
Relative atomic mass	78.96
Atomic number	34
Atomic radius	0.14 nm
Covalent radius	0.116 nm
Electronegativity (Pauling's)	2.55
Electron structure	$[Ar]3d^{10}4s^24p^4$
Oxidation states	-2, 0, +2, +4, +6
Stable isotopes	
Mass	74 76 77 78 80 82
Natural abundance (%)	0.87 9.02 7.85 23.52 49.82 9.19

[a] From: Rosenfeld & Beath (1964) and Cooper et al. (1974).

All of the oxidation states of the element listed in Table 1 are commonly found in nature except the +2 state. However, selenium compounds containing the divalent positive ion are known.

The natural isotopic pattern has been useful in identifying selenium-containing molecular fragments produced in mass spectrometry. While there are no naturally occurring radio-isotopes of selenium, several can be produced by neutron activation. Of these, ^{75}Se has the longest half-life (120 days) and is used as a tracer in experiments as well as in the determination of selenium by neutron activation analysis. Two relatively short-lived radioisotopes, [77m]Se (17.5-second half-life) and [81]Se (18.6 min half-life) have also been used in neutron activation analysis (Heath, 1969-70; Alcino & Kowald, 1973).

Selenium in the +6 or selenate state is stable under both alkaline and oxidizing conditions. Thus, it occurs in alkaline soils, where it is soluble and easily available to plants. It is also the most common form of the element found in alkaline waters. Because of its stability and solubility, it may be potentially the most environmentally dangerous form of the element.

Selenium in the +4 state occurs naturally as selenite. In alkaline solution, it tends to oxidize slowly to the +6 state, if oxygen is present, but not in an acid medium. It is readily reduced to elemental selenium by a number of reducing reagents, ascorbic acid or sulfur dioxide being commonly used for this purpose. It readily reacts with certain o-diamines, and this is used as the basis for some analytical methods. Selenium dioxide, the anhydride of selenious acid, sublimes at 317 ˇC. This is important with regard to air pollution through the combustion of materials containing the element (Heath, 1969-70; US NAS/NRC, 1976), and also to air sampling procedures.

Selenite binds tightly to iron and aluminium oxides. Thus, it is quite insoluble in soils and generally not present in waters in any appreciable amount. The nature of this binding has been suggested to be (Howard, 1971):

(a) hydroxylation of fracture surfaces of oxides in an aqueous environment;

(b) development of a pH-dependent charge on the surface by amphoteric dissociation of the hydroxyl groups;

(c) electrostatic attraction of ions of negative charge (biselenite, for instance) to the surface when it is positively charged; and

(d) specific adsorption of the ion through exchange with surface hydroxyl groups.

Elemental selenium (Se^O), like elemental sulfur, exists in several allotropic forms. At the molecular level, while several Se aggregates can form, only rings containing 8 selenium atoms (Se^8) and Se_n polymeric chains exist at room temperature. Se^8 is soluble in carbon disulfide, but Se_n is not. Both forms have a very low order of solubility in water and dilute acids or bases. Se^O exists in both amorphous and crystalline forms. Colloidal Se^O, prepared by reducing aqueous solutions of selenite, is an amorphous form with a reddish colour, used in some early methods for determining selenium at low concentrations. On heating at 60 ˇC, for a short time, colloidal selenium crystallizes and its colour changes to black. Elemental selenium boils at 684 ˇC, and since it, as well as hydrogen selenide, may form during the pyrolysis of organic materials, the use of dry ashing in preparing samples for analysis has serious limitations. This property also contributes to the problem of

atmospheric contamination with selenium in certain industrial
processes (Crystal, 1973). Elemental selenium is very stable
and highly insoluble. It is formed on the reduction of
selenate as well as of selenite. The stability and insolubility of elemental selenium render it unavailable to plants.
Its formation by natural processes might, thus, be considered
one means by which the element is removed from active cycling
in the environment.

In its -2 state, selenium exists as hydrogen selenide and
in a number of metallic selenides. Hydrogen selenide is a
strong reducing agent and a relatively strong acid with a
pK_a of 3.73. It is a gas at room temperature and very
toxic. Exposure to hydrogen selenide results in olfactory
fatigue, and individuals exposed to it may soon be unaware of
its presence. In air, it decomposes rapidly to form Se^o and
water. Selenides of heavy metals occur naturally in many
minerals, and iron selenide may be one of the insoluble forms
of the element in soils (Sindeeva, 1959; Nazarenko & Ermakov,
1971; Johnson, 1976; US NAS/NRC, 1976).

A large number of selenium analogues of organic sulfur
compounds are known. Many have been identified in plants,
animals, and microorganisms. However, although some aspects
of the metabolism of selenium resemble those of sulfur, in
many cases, their metabolic pathways diverge considerably
(Levander, 1976a). As in the case of sulfur, many of the
selenium compounds are odoriferous, volatile, and relatively
unstable. Because of this volatility, precautions are necessary in handling some samples for analysis (Klayman & Gunther,
1973; Irgolic & Kudchadker, 1974).

2.2 Analytical Methods

2.2.1 Sample collection, processing, and storage

Initially, attention must be given to the adequacy of the
method used to analyse for selenium. It is strongly recommended that the analytical method under consideration be
verified against samples of known certified selenium content
such as the standard reference materials available commer-
cially from the US National Bureau of Standards. Moreover,
adequate analytical quality control measures should be
standard operating practice in any laboratory conducting
analyses for selenium. Equally important is the proper
collection and treatment of samples before analysis. The
sample collected must be representative of the material being
studied, and must be protected from either contamination or
loss of selenium during analysis. These considerations are
especially important in the processing of organic matter
containing selenium and in the determination of trace amounts

of selenium in materials of environmental interest such as soils, air, and water.

In the case of soils, it should be recognized that sampling to a depth of 1 m is more meaningful than shallower sampling, especially where soil-selenium levels are excessively high (Olson et al., 1942). Also, considerable variations will occur in soils lying only a few metres from each other. In air samples, both volatile and particulate selenium may be present and a dry filter with an added liquid filter has been suggested for measuring these two forms and the total selenium in air (Diplock et al., 1973; Olson, 1976). Water samples may contain suspended solids and, in sampling for analysis, it must be decided whether these solids, which may contribute significant amounts of selenium, should be included in, or excluded from, the sample. Animals are known to synthesize volatile selenium compounds, and so it is preferable to analyse animal specimens without drying.

2.2.2 Sample decomposition or other preliminary treatment

Some methods of analysis can be performed without destroying the sample. In others, the sample must be treated in some way to remove organic matter, release the selenium, and bring it to the proper oxidation state. Several procedures for accomplishing this have been developed and are discussed below.

2.2.2.1 Wet digestion

Wet digestion is most commonly used for freeing the selenium and destroying the organic matter. It can be used for a wide variety of materials, wet or dry, and with many procedures for measuring the selenium. Different mixtures of nitric, sulfuric, and perchloric acids, with or without such additives as hydrogen peroxide, mercury, molybdenum, vanadium, or persulfate, have all been used with equal success (Olson et al., 1973; Olson, 1976). Some analysts find it convenient to add nitric acid alone to their samples and allow them to stand overnight at room temperature. This procedure facilitates the later steps in the wet digestion and reduces the likelihood of foaming and/or clarring of the sample. However, reproducible recoveries of selenium are obtained only if all traces of nitric acid are removed from the sample digests by heating until the appearance of perchloric acid fumes for 15 - 20 min (Haddad & Smythe, 1974). The trimethylselenonium ion of urine is somewhat resistant to decomposition by wet digestion, so an extended period of digestion is recommended for both urine and certain plant materials that may contain the trimethyl-selenonium moiety (Olson et al., 1975). While its presence in significant amounts has not been demonstrated in most tissues, the trimethylselenonium moiety may occur in kidney in such an amount that extended digestion of this tissue may also be wise. Wet digestion readily adapts to the predigestion procedure mentioned below and it can be adapted for the handling of large numbers of samples (Chan, 1976; Whetter & Ullrey, 1978). Special care is needed to avoid explosions when using perchloric acid, including the use of a digestion rack with a fume manifold and a special fume hood. However, a wet digestion technique with phosphoric acid, nitric acid, and hydrogen peroxide has been described that avoids the use of perchloric acid (Reamer & Veillon, 1983a). This procedure gives results for plasma and urine samples similar to those obtained with the traditional fluorometric method using perchloric acid digestion. However, the method needs to be validated for other sample matrices.

2.2.2.2 Predigestion

A predigestion process can often reduce sampling error, when solubilized samples or wet digestion are used in determining selenium. A sample much larger than that required for the actual determination is heated with about 10 volumes of concentrated nitric acid at a slow boil for about 20 min. After cooling and making to volume with water, an appropriate aliquot can be removed for analysis. With samples of high fat content, the fat will rise to the surface on cooling and it can be avoided in sampling for analysis. The fat contains essentially no selenium at this point. Feed samples fortified by the addition of sodium selenite or selenate, as well as premixes containing these compounds, can often be handled much more successfully with this procedure than by relying on fine grinding to assure a representative sample (Olson, 1976).

2.2.2.3 Combustion

Open combustion, with or without added fixatives, and low temperature ashing with excited oxygen have not been successfully used in selenium determination. Using the Schoeniger flask for destroying organic matter and oxidizing the selenium gives excellent results, but the method is somewhat inconvenient, and this has limited its use (Olson et al., 1973).

2.2.2.4 Fusion

Sodium carbonate or sodium peroxide fusion of soils or rocks and the Parr bomb fusion of some organic materials have given satisfactory results in selenium determination. However, these methods are inconvenient in many respects and are little used (Olson et al., 1973).

2.2.2.5 Concentration

Materials containing very small amounts of selenium may require a step for concentrating the element. With water, this can usually be accomplished by evaporation under alkaline conditions. Other methods of concentrating selenium include coprecipitation with iron hydroxide (selenite) and some of the techniques used for separating the element from interfering substances, which will be discussed in section 2.2.3.

2.2.3 Removal from interfering substances

In many methods, it is necessary to isolate the selenium or to remove certain interfering substances before the selenium is measured. Procedures used include: coprecipitation with arsenic, tellurium, or ferric hydroxide; ion exchange column chromatography; solvent extraction of selenium halides, of organic selenium complexes, or of certain interfering metals; distillation of the tetrabromide; volatilization as hydrogen selenide; precipitation as elemental selenium; paper, thin-layer, or gas chromatography; and the ring oven technique (Alcino & Kowald, 1973; Cooper, 1974; Olson, 1976).

Interference by certain metal ions has been overcome, to a large extent, by the addition of complexing reagents such as oxalate or ethylenediaminetetraacetic acid.

2.2.4 Detection and identification of selenium

Selenium in the form of selenite or selenate can be detected and identified in a number of ways (Alcino & Kowald, 1973). Such qualitative tests may be useful in some industrial situations, but normally quantitative analysis is required for confidence in any interpretation, and in the end may only be slightly more time-consuming.

2.2.5 Measurement of selenium

Several reviews of methods for measuring selenium in a wide variety of materials (Nazarenko & Ermakov, 1971; Alcino & Kowald, 1973; Olson et al., 1973; Cooper, 1974; Olson 1976) have been used as a basis for the following discussion.

The method used will depend on the sample to be analysed, the sensitivity required, the accuracy required, other analyses to be made, the number of samples, and the equipment and expertise available.

For samples containing high concentrations of the element, gravimetric or titrimetric methods may be the best. The gravimetric methods have been based on precipitation of selenium as the element, the sulfide, the salts of certain heavy metals, an organic complex, or electrogravimetrically as Cu_2Se. Titrations have been based on iodometry, argentometry, potassium permanganate reductions by selenite, or backtitrations using thiourea or sodium thiosulfate.

Other methods, many of them very sensitive, can be classified as: colorimetric, spectrophotometric, fluorometric, atomic absorption spectrometric, X-ray fluorescent, gas chromatographic, neutron activation, proton activation, polarographic, coulometric, potentiometric, spark source mass spectrometric, gas chromatograpic, mass spectrometric, anodic stripping voltammetric, and catalytic.

At one time, the colorimetric determination was the most used for the analysis of samples containing small amounts of selenium. The method most commonly used (Robinson, 1933) was specific for selenium and, for its time, quite sensitive. The titrimetric method of Klein (1943) superceded it, being somewhat more reproducible and sensitive, until selenium was found to have a role as an animal nutrient, when considerably greater sensitivity was required. Very sensitive methods, most commonly used, include fluorometry, neutron activation, and atomic absorption spectrophotometry.

2.2.5.1 Fluorometric analysis

The methods most widely used for the determination of selenium in natural materials are based on fluorometry. Selenious acid reacts with l-diamines to give a piazselenol, which is fluorescent. The l-diamine of choice is 2,3-diaminonaphthalene. The piazselenol is extracted from an acid solution (pH 1 - 2) with either decahydronaphthalene or cyclohexane, in either of which the fluorescence yield is good. When the organic matter is destroyed, the reaction can be specific for selenium. Interference from copper, iron, and vanadium, and some oxidizing agents can usually be avoided.

Fluorometric methods have been applied to a number of materials; those for foods and plants have been subjected to collaborative studies (Olson, 1969; Ihnat, 1974) and have been accepted as official first, and final action, by the Association of Official Analytical Chemists (AOAC, 1975). Critical reviews of fluorometric methods (Haddad & Smythe, 1974; Michie et al., 1978) have shown them to be reliable providing that the precautions described for various adaptations are followed. The methods are sensitive to about 0.01 μg of selenium in a sample, are adaptable to large numbers of samples, can be automated (Brown & Watkinson, 1977; Szydlowski

& Dunmire, 1979; Watkinson, 1979; Watkinson & Brown, 1979) and can be simplified (Spallholz et al., 1978; Whetter & Ullrey, 1978). While isotope dilution procedures can be used with fluorometric methods (Cukor et al., 1964), they are not essential for reliable results. Fluorometric analyses rely on perchloric acid digestion of the samples, and, thus, require a perchloric acid fume hood. Furthermore, 2,3-diaminonaphthalene of adequate purity is sometimes difficult to obtain. Improved sensitivity of the fluorometric analysis of water and biological samples was achieved by redistilling cyclohexane and extracting 2,3-diaminonaphthalene solution in 0.1 M hydrochloric acid with two portions of redistilled cyclohexane (Nazarenko et al., 1975; Nazarenko & Kislova, 1977). Wilkie & Young (1970) have also published detailed experimental procedures for purifying various reagents used in fluorometric analysis. The analysis can be carried out with a simple fluorometer, is adaptable to a wide variety of materials, is reasonably rapid and reliable, and is relatively inexpensive to perform.

2.2.5.2 Neutron activation analysis

Thermal neutron activation is the most commonly used procedure for irradiating samples containing selenium. Of the radionuclides of selenium that it produces, ^{75}Se (half-life 120 days), ^{81}Se (half-life 18.6 min), and ^{77m}Se (half-life 17.5 seconds) have been useful in analysis. $^{81\ m}$Se (half-life 57 min) has been used to measure carrier selenium after separation and reirradiation (Kronborg & Steinnes, 1975). Proton-induced X-ray fluorescence methods have also been reported (Bearse et al., 1974; Barrette et al., 1976).

Two methods of converting radioactivity measurements to selenium content are:

(a) the direct method using calculations based on certain known values and constants; and

(b) the comparator method, based on the irradiation and counting of a known amount of selenium, as a standard, together with the samples. The comparator method is the most accurate and most often used, but the direct method is often used when short-lived isotopes are measured. Ruthenium has been recommended as a multi-isotopic comparator when multi-element analysis is being performed (Van der Linden et al., 1974).

Neutron activation analysis can be used without and with sample destruction. When ^{75}Se is measured by non-destructive analysis, a period of several weeks or months may be allowed for decay of some interfering radioisotopes. The ^{77m}Se isotope has been used without this type of delay with some success. Non-destructive methods have been used for multi-element analysis, but they have been subject to more errors, because of interference, than destructive methods followed by chemical separation. The use of high resolution gamma-ray spectrometry substantially increases the specificity of the non-destructive methods and the complete processing of 300 - 400 samples per month is possible (Pelekis et al., 1975). A method has been developed using irradiation with epithermal neutrons and multiparameter analysis which eliminates the effect of spurious interferences in the determination of low selenium concentrations by instrumental neutron activation analysis. High specificity has been achieved by the selective response of the system to gamma-gamma coincidences of ^{75}Se (Vobecky et al., 1977, 1979; Pavlik et al., 1979).

Destructive neutron activation analysis is used with the chemical separation of the selenium. In most methods, ^{75}Se is measured, but ^{77m}Se measurement is not uncommon, and ^{81}Se measurement has also been used. Carrier selenium is usually added following the irradiation. Destruction of the sample has been accomplished by wet digestion, sodium peroxide fusion, or combustion. Combustion has been accomplished in a closed system, which provides for the dry distillation of selenium and certain other elements into a trapping system. Wet digestion is normally followed by separation of the element by distillation as the tetrabromide, and/or its precipitation, adsorption on an ion exchange material, solvent extraction of an organic selenium complex, or reversed phase extraction chromatography. A semi-automated method based on wet digestion, separation by distillation, and the use of ruthenium as a comparator has been reported (D'Hondt et al., 1977).

Neutron activation analysis can be very accurate, sensitive, and specific, especially when used with sample destruction and chemical separation of the selenium. It adapts well to multi-element analysis. However, it requires sophisticated equipment that most laboratories do not have. Furthermore, it is time-consuming, especially when used with chemical separation. Its most important use may be as a reference method against which other methods may be evaluated or where good reagents are not available.

2.2.5.3 Atomic absorption spectrometry

Methods for selenium analysis now being most actively studied are those based on atomic absorption spectrometry

(AAS). A wide variety of techniques has been described. Most require some type of wet digestion, when organic matter is present. Flame atomization methods are useful for materials with a high selenium content. However, other more sensitive methods have been developed, the most used being based on hydrogen selenide generation. This has the advantages of separating the selenium from many other elements and measuring it using techniques that provide excellent sensitivity. Ihnat (1976) compared the performance of hydride generation with that of a carbon furnace atomization-AAS method, finding hydrogen selenide generation to be superior. In further studies, Ihnat & Miller (1977a,b) found that the sensitivity of this method was excellent and its accuracy, fairly good, but that its precision in a collaborative study was not. McDaniel et al. (1976) found that procedures for hydrogen selenide generation might liberate as little as 10% of the element from solution. They suggested means for optimizing hydride generation to give a simple and sensitive selenium determination, free from interferences, and using a heated graphite atomizer for the measurement of the element. The hydride generation technique has been automated (Goulden & Brooksbank, 1974; Pierce & Brown, 1977; Pyen & Fishman, 1978), and improved equipment for hydride generation has recently been described (Brodie, 1979). Encouraging results with graphite furnace atomization in the presence of nickel (Shum et al., 1977) indicate that this method may also find considerable use.

Atomic absorption methods based on Zeeman effect background correction to remove spectral interferences (Pleban et al., 1982; Carnrick et al., 1983) are currently receiving attention. Such procedures offer the promise of carrying out selenium analyses in certain biological matrices directly, i.e., without the need for sample digestion. Development of such techniques would greatly increase the speed and convenience of selenium analysis.

2.2.5.4 Other methods

Examples of other methods that have received recent attention and may find some more use in the future include: energy dispersive X-ray fluorescence (Holynska & Markowicz, 1977); differential pulse polarography (Bound & Forbes, 1978); anodic (Andrews & Johnson, 1976) or cathodic (Blades et al., 1976) stripping voltammetry; atomic fluorescence spectrometry (Thompson, 1975); gas-liquid chromatography (Ermakov, 1975; Poole et al., 1977); and spark source mass spectrometric isotope dilution (Paulsen, 1977). A gas chromatographic, mass spectrometric double isotope dilution technique based on a volatile chelate of selenium has recently been described, which not only determines total selenium in biological materials but also allows stable selenium isotopes to be followed as tracers in metabolic studies (Reamer & Veillon, 1983b). A variety of methods has been proposed and used for identifying and measuring a number of chemical forms of selenium in different materials including: paper (Hamilton, 1975) or ion exchange (Martin & Gerlach, 1969) chromatography for selenium compounds in plants; gas chromatography for volatile selenium compounds (Doran, 1976); and ion exchange (Shrift & Virupaksha, 1965) or solvent extraction separation (Kamada et al., 1978) for measuring selenite and selenate in solutions.

3. SOURCES, TRANSPORT, AND CYCLING OF SELENIUM
IN THE ENVIRONMENT

3.1 Natural Sources

Because selenium is present in natural materials, its
occurrence in any substance cannot be assumed to be the result
of human activity and some knowledge of its natural
distribution is required before evaluating its role as a
pollutant.

3.1.1 Rocks and soils

Several reviews of selenium geochemistry have been published (Sindeeva, 1959; Rosenfeld & Beath, 1964; Lakin &
Davidson, 1967; Cooper et al., 1974). The concentration of
selenium in ingneous rocks is low, usually much less than 1
mg/kg, and similar levels probably occur in metamorphic rocks.
Sedimentary rocks, such as sandstone, limestone, phosphorite,
and shales may contain from < 1 to > 100 mg/kg.

The selenium content of a soil reflects, to some extent,
that of the parent material from which the soil has been
formed. Thus, in arid and semi-arid areas, soils of high
selenium content have been derived from sedimentary rocks,
usually shales and chalks (Moxon et al., 1950). These soils
are alkaline in reaction, favouring the formation of selenate
(Geering et al., 1968), which is readily available to plants
(Moxon et al., 1950). The selenate is easily leached from the
surface soil, but, with limited rainfall, it is redeposited in
the subsurface soil, where it is still available to plants
(Olson et al., 1942). Thus, surface soil analysis has not
been found to be a reliable measure of the potential of a soil
to produce vegetation containing toxic levels of selenium.

Coal with an unusually high selenium content (> 80 000
mg/kg; average 300 mg/kg) was recently identified as the
ultimate environmental source of selenium contaminating soils
in a seleniferous region of Enshi county, Hubei province, the
People's Republic of China (Yang et al., 1983). It was
thought that selenium passed from the coal to the soil through
weathering, leaching, and possibly biological action, thus
making it available to the crops. Lime fertilizers,
traditionally used in the area, would also render the selenium
accumulated in the soil more readily available to plants.

Some soils produce crops nutritionally deficient in
selenium (US NAS/NRC, 1971). The most obvious factor in the
formation of these soils is probably the parent material
(Hodder & Watkinson, 1976), but other factors such as rainfall, climate, pH, and soil composition contribute signifi-

cantly (Gissel-Nielsen, 1976, 1986). As with producing plants
of high selenium content, soils producing plants low in the
element cannot be identified by analysis of the surface soil
alone (Shacklette et al., 1974). Because of the many factors
affecting availability, plant analysis is also necessary
(Kubota et al., 1967).

Sun et al. (1985) reported that the average total selenium
content (112 µg/kg, range 59 - 190 µg/kg) in the soil of 6
low-selenium Keshan disease areas in China was significantly
lower than that of the 5 corresponding non-endemic areas (234
µg/kg, range 142 - 318 µg/kg). The average water-soluble
selenium content and the percentage of water-soluble selenium
in the total soil-selenium of the endemic areas (4.0 µg/kg,
range 2.2 - 8.7 µg/kg and 39 g/kg, respectively) were also
significantly lower than those of the non-endemic areas
(19.9 µg/kg, range 11.4 - 38.8 µg/kg and 9.2 g/kg, respectively). The total soil-selenium content of a high-selenium
area in China was 7865 µg/kg (range 6390 - 10 660 µg/kg),
but the percentage of water-soluble selenium was not very high
(30 - 40 g/kg).

3.1.2   Natural selenium in the food chain

All plants absorb selenium from the soil, the amount
depending mainly on the species, the stage of growth, and the
availability of the selenium in the soil. Biogeochemical
factors influencing the availability of selenium in the soil,
such as pH, iron content, etc., have been reviewed by Ermakov
& Kovalskij (1968), Allaway (1973), and Kovalskij (1974,
1978). Various plant tissues contain different selenium
levels, generally following protein content. Certain plants,
known as selenium accumulator plants, take up quantities of
selenium great enough to be toxic for animals. The selenium
content of animal tissues reflects that of the feeds consumed
(Allaway, 1973; Kovalskij & Ermakov, 1975). Thus, naturally-
occurring selenium readily passes up through the food chain
via animals to human beings, and the amount of selenium in the
human diet is largely determined by the amount of selenium in
the soil available for absorption by plants. However, as
discussed in section 7, the increased retention of selenium by
animals, under conditions of low selenium intake, and the
decreased retention during high selenium intake ensures that,
in different geographical zones, the range of selenium levels
in animal products is less than that in plant tissues. This
"buffering effect" of animals in the food chain tends to
moderate the extremes of selenium intake to which human beings
are exposed, whereas herbivores are subject to much greater
fluctuations in selenium intake.

3.1.3   Water and air

Under natural conditions, the concentration of selenium in water usually ranges from a few tenths to 2 or 3 µg/litre (Ermakov & Kovalskij, 1974; US NAS/NRC, 1976). The highest natural concentration reported to date is 9000 µg/litre, almost all other values falling below 500 µg/litre (US NAS/NRC, 1976). A garlicky odour has been noted in waters containing 10 - 25 µg selenium/litre and an astringent taste can be detected in water samples containing 100 - 200 µg/litre (Pletnikova, 1970). In a study carried out in Nebraska, USA, which was biased towards finding waters with elevated selenium contents, it was reported that about one-third of 161 well samples contained over 10 µg selenium/litre and about 4%, over 100 µg/litre (Engberg, 1973). Surface waters seem much less likely to contain excessive levels of selenium than ground waters.

Soils, plants, microorganisms, animals, and volcanoes all contribute selenium to the atmosphere. All of these sources, and possibly also certain sediments, produce volatile forms, and soils and volcanoes probably contribute particulate matter containing the element. Some man-made activities also yield atmospheric selenium (section 3.2.2), and it is difficult to establish the proportion of selenium that comes from these activities and the proportion that comes from natural sources (US NAS/NRC, 1976). However, data concerning atmospheric selenium over the poles and over the Atlantic Ocean, as well as over areas of minimal human activity (Zoller et al., 1974; Duce et al., 1975; Dams & De Jonge, 1976) suggest that the average concentration from natural sources should be less than 0.04 ng/m³, except close to centres of volcanic activity.

3.2   Man-Made Sources

3.2.1   Agriculture

Early agricultural uses of selenium compounds as pesticides were very limited and short-lived. The recent use of selenium compounds as feed additives or injectables for the prevention of selenium deficiency diseases in farm animals represents a source for environmental contamination, but, compared with the levels already present in most feeds and the amounts found in soils, even in the deficient areas, this source seems insignificant (US NAS/NRC, 1976). In New Zealand, little increase in the selenium content of human foods was observed after the introduction of selenium supplementation for farm animals (Thomson & Robinson, 1980). The use of fertilizers (Koivistoinen & Huttunen, 1985) and foliar sprays (Gissel-Nielsen, 1973, 1986) to rectify selenium

deficiency in feeds is being introduced, but they are unlikely
to become a hazard in the environment. Other fertilizers do
not contain enough selenium to contribute significantly to the
environment (Gissel-Nielsen, 1971). Irrigation of seleniferous lands may result in an increased concentration of
selenium in drainage waters (US NAS/NRC, 1976), but until
recently there was no evidence to suggest that this constitutes a hazard. However, in the San Joaquin Valley of
California, reproductive problems have now been observed in
aquatic birds that used irrigation drain-water ponds as
waterfowl habitat (Ohlendorf et al., 1986). Although the
toxic signs reported in these birds resembled avian selenosis,
the possible role of other water-borne toxins such as high
levels of arsenic or boron has been pointed out.

3.2.2 Industry

The main industrial sources emitting selenium into the
environment include the mining, milling, smelting, and
refining of copper, lead, zinc, phosphate, and uranium, the
recovery and purification of selenium itself, the use of
selenium in the manufacture of various products, and the
burning of fossil fuels. Some problems from selenium
emissions might arise in regions near selenium-producing or
coal-burning industries. For instance, in areas where copper
sulfide ores were mined and processed, atmospheric selenium
levels ranging from 0.15 to 6.5 $\mu g/m^3$ were reported within
0.5 - 10 km of the ore-processing plant (Seljankina et al.,
1974). Selenium pollution of the air may be important with
regard to possible contamination of open-air water reservoirs;
waterways can be contaminated directly by selenium from mining
or industrial effluents. For example, waste-waters of ore
mines and effluents from a number of non-ferrous industries
were reported to contain selenium in the concentration range
of 14 - 56 $\mu g/litre$. Residual selenium impurities in
conditionally cleaned effluents released into open water
bodies can contribute some contamination (Shumaev et al.,
1976). Effluents from sewage plants can add selenium to
waters; raw sewage has been reported to contain up to 280 μg
selenium/litre, and levels of 45 - 50 $\mu g/litre$ have been
reported in primary and secondary sewage effluents (Baird et
al., 1972). Emission factors for atmospheric selenium
contamination and solid waste generation for various
industrial processes in the USA are shown in Table 2. The
pattern of such emissions will vary depending on the
industrial and mining characteristics of individual
countries. In Canada, for example, over half of the air
pollution due to selenium resulted from primary copper and

Table 2. Estimated selenium materials balance for the USA for 1970[a],[b]

Source	Selenium input	Atmospheric emissions	Solid waste
	(--------tonne---------)		
Industrial production			
Mining and milling	2900	5	1633
Smelting and refining	1270	227	181
Selenium refining	877	59	362
Industrial consumption			
Glass and ceramics	59	59	
Electronics and duplicating	227	[c]	[c]
Pigments	59	[c]	[c]
Iron and steel alloys	23	11	[c]
Other	38	[c]	[c]
Other			
Coal consumption	1315	680	635
Fuel oil consumption	59	59	0
Incineration	36	[c]	36
Other disposal	322	[c]	318
Total emmissions:		1100	3165

[a] From: US NAS/NRC (1976).
[b] The input not accounted for as atmospheric emissions or solid waste would appear in intermediate or commercial products.
[c] Less than 1 tonne.

nickel production, the combustion of coal contributing less than 25%.

When coal from deposits throughout the USA was analysed (138 analyses), the highest selenium content was 10.65 mg/kg and the average, was 2.8 mg/kg, a hundred times less than the high-selenium coals reported in China (Yang et al., 1983) (section 3.1.1). The chemical form of the element in coal is not known (US NAS/NRC, 1976). The selenium from the coal was released into the surrounding soils by natural processes, thereby contaminating the food chain with toxic levels of selenium.

Few data on the selenium content of fuel oils are available. Most reported values have been below 1 mg/kg (US NAS/NRC, 1976). Oil shale has been reported to contain selenium (Shendrikar & Faudel, 1978), but apparently there are no reports of its occurrence in natural gas.

3.3 Environmental Transport

Because of the lack of quantitative data, it is impossible to evaluate the relative importance of various processes in the environmental transport of selenium. However, Bertine & Goldberg (1971) have estimated that rivers move over 7 x 10^6 kg of selenium every year, even though surface waters generally contain little of the element. The quantity of selenium passing via grain into the world feed and food supplies would be at least 10^5 kg, if it is assumed that the annual global production of cereal grains and corn is 10^9 tonnes and that these grains contain an average of 0.1 mg selenium/kg. Although this simple calculation neglects all other plants and the large ocean flora, the quantity of selenium moved in grains alone is equivalent to about 7% of the average world production of the element for industrial purposes. Thus, natural geophysical and biological processes probably play dominant roles in shaping the present status of selenium in the human environment and must be taken into account in any evaluation of the effects of man's activities on selenium in the environment.

3.4 Biological Selenium Cycle

Several diagrammatic schemes have been presented for the cycling of selenium in nature (Moxon et al., 1950; Lakin & Davidson, 1967; Olson, 1967; Frost & Ingvoldstad, 1975; US NAS/NRC, 1976). None of the schemes is complete and none provides good quantification for the various steps in the cycle. Indeed, a complete scheme is perhaps too complex to be presented diagrammatically with clarity, and reasonably accurate quantification of the various steps would demand much more background information than is available at present. It appears that selenium has many pathways for distribution among geological media, biological media, the atmosphere, and waters by geophysical, biological, and man-made activities. It also appears that there are "sinks" into which selenium may fall (elemental selenium, ferric hydroxy complexes, and metal selenides) from which it may be only very slowly recycled. These may represent a natural detoxification process. However, the fact that there are areas of deficiency and areas of excess of the element over the earth's surface is evidence that its depletion or concentration can occur locally. How quickly changes in the selenium status of a region occur is not known, and there is a distinct need for data from which an evaluation could be made of the rate and impact of these changes as well as the influence of man's activities on them. For example, Allaway (1973) concluded that any soil-plant-animal chain of food production on neutral or acid soils would eventually become depleted in biologically effective selenium, though it is not known how rapidly this might occur. Shrift (1973) suggested the existence of a biological selenium cycle from the standpoint of transformations between several oxidation-reduction states, as shown in Fig. 1. Reduction of the element is common and well established, but its biological oxidation is less well documented. However, a missing link on the oxidative side of the biological selenium cycle was found by Sarathchandra & Watkinson (1981), when they discovered a strain of <u>Bacillus megaterium</u>, isolated from soil, that oxidized elemental selenium to selenite and traces of selenate. The role of non-biological oxidation and reduction of the element has not yet been elucidated. While a balance may be maintained in the redox state of selenium throughout the world, there are regions where oxidation prevails and others where reduction prevails. Knowledge of the conditions and mechanisms leading to each situation is essential to understand the development of regions of selenium excess or deficiency.

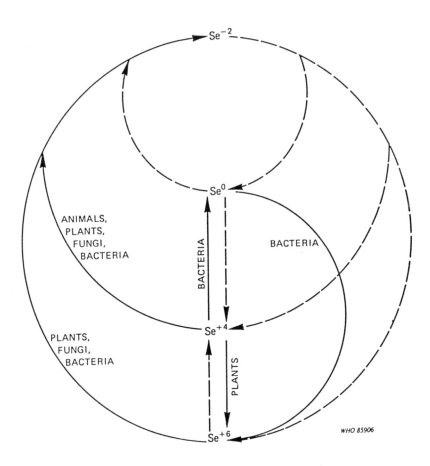

Fig. 1. Biological selenium cycle. Well-established pathways are indicated by solid lines, those needing further substantiation by dotted lines. Adapted from: Shrift (1964) and slightly modified according to the reports of Asher et al. (1977) and Sarathchandra & Watkinson (1981).

4. LEVELS IN ENVIRONMENTAL MEDIA

4.1 Levels and Chemical Forms of Selenium in Food

4.1.1 Levels in food

In spite of the relatively few data available, some generalizations concerning the selenium content of foodstuffs can be made. For example, the level of selenium in food depends on natural differences among food commodities and the natural availability of selenium in the environment. Moreover, certain of man's activities can influence the selenium content of human foods.

4.1.1.1 Natural differences among food commodities

A wide range of values has been reported in the selenium content of foods (mg/kg wet weight): liver, kidney, and seafood, 0.4 - 1.5; muscle meats, 0.1 - 0.4; cereal and cereal products, < 0.1 - > 0.8; diary products, < 0.1 - 0.3; and fruits and vegetables, < 0.1) (Oelschlager & Menke, 1969; Morris & Levander, 1970; Schroeder et al., 1970; Suchkov, 1971; Arthur, 1972; Millar & Sheppard, 1972; Ferretti & Levander, 1974; Sakurai & Tsuchiya, 1975; Abutalybov et al., 1976; Bieri & Ahmad, 1976; Kasimov et al., 1976; Olson et al., 1978). It should be pointed out that these values are given for raw foods (food as purchased) rather than cooked foods (food as eaten). The effects of cooking on the selenium content of foods are described in section 4.1.1.3. Moreover, the values presented in these food composition studies should not be compared from one country to another, because of the variations in the analytical methods and sampling procedures used (section 2).

Organ meats, such as kidneys or liver, contain the highest levels of selenium, but some sea-food products contain almost as much. Muscle meats are significant sources of selenium, though they do not contain as much as organ meats or seafoods. Certain semolina, grain, and cereal products can contribute appreciably to the dietary-selenium intake, but wide variations in selenium content have been found in different samples of the same foodstuff. Such variation is typical of many plant foods and the reasons for this are discussed below. Milk, cheese, and egg samples from several countries showed low to moderate values for selenium, but again the results were quite variable. Fruits and vegetables generally contained very low levels of selenium, though garlic and mushrooms contained moderate levels of the element.

The selenium contents of baby foods tended to show the
same general pattern as those in adult foods (Morris &
Levander, 1970; Arthur, 1972), i.e., meat and cereal products
contained the highest levels and fruit and vegetable products,
the lowest. Shearer & Hadjimarkos (1975) analysed samples of
mature human milk from 241 subjects living in the USA and
found that the overall mean selenium content was 0.018
mg/litre. Over 98% of the samples contained between 0.007 and
0.033 mg/litre. Grimanis et al. (1978) reported an average of
0.015 mg/litre in 5 samples of mature human milk collected in
Greece. These authors found slightly higher values in 15
samples of human transitional milk (0.016 mg/litre) and
colostrum (0.048 mg/litre). A slightly lower average selenium
concentration of 0.013 mg/litre, was reported for samples of
human transitional milk in New Zealand (Millar & Sheppard,
1972). The most extreme values for the selenium content of
human milk were reported from China and ranged from 0.0026
mg/litre in areas where Keshan disease was prevalent to 0.283
mg/litre in high-selenium areas (Yang et al., 1986).

It can be seen from Table 3 that meat-based infant
formulae have a higher selenium content than formulae based on
milk or soy protein, and that casein-based powdered formulae
for special medical purposes (diet therapy for errors in amino
acid metabolism or malabsorptive states) contain low levels of
selenium. Blended food tube-feeding formulae containing meat
tend to have higher selenium contents than comparable products
containing only milk or casein and soy protein (Table 3).
Chemically-defined diets having egg albumen as the protein
source contained more selenium than diets based on casein
hydrolysate, which, in turn, contained more selenium than
diets based on purified amino acids. Total parenteral nutrition solutions based on casein hydrolysate also contain more
selenium than solutions based on amino acid mixtures, which
contain very low levels of selenium.

One of the factors that can influence the amount of
selenium in plant foodstuffs is the nature of the plant
itself. Plants have been divided into 3 groups depending on
their tendency to take up selenium from seleniferous soils
(Rosenfeld & Beath, 1964):

Group 1   Primary selenium accumulators - can contain very
          high amounts of the element (often over 1000
          mg/kg dry weight).

Group 2   Secondary selenium accumulators - rarely contain
          more than a few hundred mg/kg.

Group 3   Many weeds and most crop plants, grains, and
          grasses - rarely contain more than 30 mg/kg,

Table 3. Selenium content of commercial formula diets[a]

Product	Selenium content	
	(mg/kg wet weight)	(mg/kg dry weight)
Infant formulae:		
milk-based	0.004 - 0.027	
soy-based	0.004 - 0.030	
meat-based	0.046 - 0.070	
casein-based formulae for special medical purposes (powders)		0.048 - 0.120
Food supplements and tube-feeding formulae:		
milk-based	0.005 - 0.045	
soy-casein-based	0.013 - 0.020	
blended foods	0.037 - 0.056	
Chemically defined diets: (powders)		
egg albumen low residue		0.224 - 0.351
egg albumen moderate nitrogen		0.388 - 0.886
egg albumen high nitrogen		0.503 - 0.570
casein hydrolysate		0.052 - 0.071
amino acid mixture		0.001 - 0.011
Total parenteral nutrition solutions:		
casein hydrolysate	0.037 (0.032 - 0.041)	0.324
diluted 1:1 with 50% dextrose solution	0.019 (0.017 - 0.020)	0.093
amino acid mixture	0.001	0.010

[a] Adapted from: Zabel et al. (1978).

even when grown on seleniferous soils; when grown on normal soils generally contain less than 1 mg/kg.

Plants from Groups 1 and 2 do not usually contribute to the selenium intake of human beings, since they are not consumed directly by people and are consumed by animals only when other feeds are not available. However, plants from Group 3 can contribute large amounts of selenium, if they are grown in seleniferous areas (see below).

The concentrations of many nutritionally essential trace elements are known to be decreased by the milling of grains into cereals (Czerniejewski et al., 1964) and preliminary analysis of random supermarket samples of cereal foods suggested that refined products such as white flour or white bread contained less selenium than whole grain foods such as whole wheat flour or whole wheat bread (Morris & Levander, 1970). However, a subsequent more controlled analytical study in which various grain fractions were taken from the same production batch showed that milling a variety of grains decreased the concentration of selenium in the consumer product by only 10 - 30% (Ferretti & Levander, 1974). Thus, the selenium content of cereal products is somewhat less than that of the parent grains but the decreases in concentration are not nearly as great as those observed with other essential trace minerals.

Selenium tends to be localized mainly in the protein fraction of plant and animal tissues; thus, the protein content of a food also influences its selenium content. For example, the concentration of selenium in a series of soybean products prepared from a given lot of soybeans increased as the protein content of the product increased (Ferretti & Levander, 1976). However, protein content is only an expression of the potential of a food to contain selenium, and a food high in protein will not necessarily also be high in selenium. Moreover, non-protein seleno amino acids occur in some foods and thus, the protein content may not always reflect the selenium content.

4.1.1.2 Effects of natural differences in the availability of selenium in the environment on levels in food

The most important factor in determining the selenium content of plant foods and feeds is the amount of selenium in the soil that is available for uptake by the plant. Sun et al. (1985) reported that the selenium content of local crops was significantly correlated with the selenium content of the local soil, the correlation coefficients of soybean, corn, and rice being 0.9456, 0.9953, and 0.9954, respectively. Since the water-soluble selenium in the soil was directly correlated with the pH and inversely correlated with humin and total iron in the soil, the selenium content of local crops was also influenced by the amount of water-soluble selenium in the soil. An analytical survey carried out in the USA demonstrated that the level of selenium in alfalfa plants varied from less than 0.01 to more than 5.0 mg/kg, and it was assumed that these levels reflected the amount of available selenium in the soil (Kubota et al., 1967). A variation in the selenium content of wheat from 0.04 to 21.4 mg/kg, depending on where the plant was grown, was reported by Schroeder et al. (1970). Samples of several foods bought in local markets in Caracas, Venezuela contained much more selenium than similar foods

purchased in supermarkets in the eastern USA (Table 4). The
likely source of selenium in the milk, eggs, and meat from
Caracas is sesame cake, since sesame is produced mostly in the
seleniferous area of Venezuela and the pressed cake is widely
used as an ingredient in animal feed. Selenium levels as high
as 14 mg/kg in corn and 18 mg/kg in rice have also been
observed in certain food samples taken from high-selenium
regions in Venezuela (Jaffe, 1976). These concentrations of
selenium are as high as those reported in foods from seleniferous zones of the USA (Smith & Westfall, 1937).

Table 4. Comparison of selenium contents of selected foods available in
Caracas, Venezuela, and Beltsville, Maryland, USA
(mg selenium/kg wet weight)[a]

Food	Caracas	Beltsville
Powdered milk	0.417	0.169
Whole milk	0.115	0.012
American-type cheese	0.425	0.090
Swiss-type cheese	0.382	0.104
Pork	0.833	0.209
Chicken	0.702	0.106
Egg	1.520	0.116

[a] From: Mondragon & Jaffe (1976).

Great extremes in the selenium content of staple foods
have also been reported recently from China (Yang et al.,
1983). For example, samples of corn, rice, and soybeans,
taken from a high-selenium area with a history of human
intoxication reported as chronic selenosis (section 8.1.1.1)
contained average selenium levels of 8.1, 4.0, and 11.9 mg/kg,
respectively, whereas samples of the same staples collected in
a low-selenium area where Keshan disease was prevalent (a
human selenium-deficiency disease) (section 8.2.2) contained
average selenium levels of only 0.005, 0.007, and 0.010 mg/kg,
respectively (Table 5). A low selenium content in food has
also been reported in other countries with low selenium soils,
such as New Zealand and Finland. Typical ranges for selenium
contents in food in such countries include (mg/kg wet weight):
liver, kidney, and seafood, 0.09 - 0.92; muscle meats, 0.01 -
0.06; cereal and cereal products, 0.01 - 0.07; milk, < 0.01;

Table 5. Selenium contents of staple foods grown on soils in areas of China with excess, moderate, and deficient levels of selenium[a]

Place	Corn		Rice		Soybean	
	Number of samples	Se content (mg/kg)	Number of samples	Se content (mg/kg)	Number of samples	Se content (mg/kg)
High-selenium area with a history of intoxication reported as chronic selenosis	44	8.1 (0.5-28.5)[b]	22	4.0 (0.3-20.2)[b]	17	11.9 (5.0-22.2)[b]
High-selenium area reported to be without selenosis	2	0.57	2	0.97	2	0.34
Moderate-selenium-adequate area (Beijing)	82	0.036 (± 0.056)	76	0.035 (± 0.027)	31	0.069 (± 0.076)
Low-selenium area	10	0.009 (± 0.009)	32	0.022 (± 0.009)	–	–
Low-selenium area with Keshan disease	195	0.005 (± 0.003)	49	0.007 (± 0.003)	150	0.010 (± 0.008)

a Adapted from: Yang et al. (1983).
b Mean ± SD or range shown in parenthesis.

and fruits and vegetables, < 0.01 - 0.02 (Koivistoinen, 1980; Thomson & Robinson, 1980). Samples of staple foods collected in areas where soils contained intermediate levels of selenium also contained intermediate levels of selenium.

As might be expected, the selenium contents of food products of animal origin depend heavily on the amount of naturally-occurring selenium in the feed given to the animal. In the USA, it was shown that the selenium content of swine muscle was highest in areas known to have a high level of available selenium in the soil and lowest in areas in which the available soil-selenium was low (Ku et al., 1972). This indicates that selenium is readily passed up the soil-plant-animal food chain to human beings.

4.1.1.3 Man-induced changes in selenium levels in food

(a) Human activities that increase selenium levels

Perhaps the most direct way that man's activities can increase the selenium content of the food supply is the deliberate addition of selenium to the feeds of poultry and certain livestock. In some countries, this is now accepted practice to prevent the occurrence of selenium-deficiency diseases, many of which cause significant economic losses to farmers throughout the world. In the USA, for example, farmers are permitted to add 0.1 mg selenium/kg (as sodium selenite or selenate) complete feed for beef and dairy cattle, sheep, chickens, ducks, swine (0.3 mg/kg in starter and prestarter rations), and 0.2 mg/kg for turkeys (US Department of Health and Human Services, Food and Drug Administration, 1984; Subcommittee on Selenium - Committee on animal Nutrition, 1983). It has been shown that the edible tissues of poultry, swine, and sheep, fed diets fortified with inorganic selenium salts at the regulated levels, did not contain any more selenium than the tissues of animals fed diets, naturally adequate in selenium (Allaway, 1973; Ullrey et al., 1977). The homeostatic mechanisms that appear to limit the concentrations of selenium in the edible tissues of animals fed certain levels of sodium selenite are discussed further in section 6.

Finland is the first country to decide to increase the selenium content of Finnish feed and food by the addition of sodium selenate to fertilizers, to be used in the whole country at a concentration of 16 or 6 mg/kg for cereal and grassland crops, respectively (Koivistoinen & Huttunen, 1985). The manufacture of these selenized fertilizers began in the summer of 1984 and they will be used at application rates of 10 g selenium/ha per growing season. It is anticipated that the added selenium in the fertilizers will be transported to the human food chain, and the selenium level of the cereal and grassland crops will be raised to 0.1 and 0.15 - 0.20 mg/kg (dry basis), respectively.

Another potential way in which the level of selenium in the food chain might be increased by man's activities is the proposed use of selenium-bearing fly ash as a soil supplement. Furr et al. (1977) analysed cabbages grown on potted soil supplemented with several different fly ashes and found that the selenium levels in the cabbages were closely correlated with those in the respective fly ashes in which the plants were cultured. The ready bioavailability to animals of the selenium in plants grown on fly ash was demonstrated in a study (Stoewsand et al., 1978) in which Japanese quail were fed a complete diet containing 60% winter wheat that had been grown to maturity on either soil or a deep bed of fly ash. The soil contained 2.1 mg selenium/kg and the wheat grown thereon contained 0.02 mg/kg, a level typical of wheat from selenium-deficient areas. The fly ash contained 21.3 mg selenium/kg, and the wheat grown on it contained 5.7 mg/kg, a level sometimes found in wheat from naturally seleniferous areas. The levels of selenium in the tissues of the quail fed the wheat grown on fly ash were much higher than those of quail fed the wheat grown on soil (Table 6). Moreover, the selenium contents of eggs from quail fed the wheat grown on fly ash were 3.5 mg/kg in the yolk and almost 10 mg/kg in the white compared with 0.5 and 0.2 mg selenium/kg, respectively, in eggs from quail fed the control wheat. Obviously, the use of selenium-bearing fly ash as a soil supplement to provide nutritionally desirable levels of selenium in plants should be carried out with care, since inappropriate use could lead to an unwanted build-up of selenium residues in the food chain. Also, the application of fly ash to soil at rates sufficient to correct selenium deficiency in animals may damage the soil.

Table 6. Selenium in tissues of male Japanese quail fed winter wheat grown on soil or fly ash[a]

Wheat grown on:	Tissue selenium				
	brain	heart	kidney	liver	muscle
	(mg/kg dry weight)				
Soil	0.8 ± 0.1	0.7 ± 0.2	3.6 ± 0.4	1.6 ± 0.0	0.3 ± 0.2
Fly ash	3.4 ± 1.6	4.4 ± 0.7	9.5 ± 1.1	12.7 ± 2.4	4.1±0.6

[a] From: Stoewsand et al. (1978).

The possibility has been raised that some selenium may be added to the food supply as a result of atmospheric contaminants from the burning of fossil fuel or industrial emissions settling out on the leaves of plants or on the surface of the soil. The geographical pattern of selenium levels in rainwater from Denmark or the USA implicates airborne selenium from the industrial and domestic uses of fossil fuel as sources (Kubota et al., 1975). Total industrial emissions of selenium in the USA for the year 1970 were estimated to be about 1.1 million kg, 62% of which was derived from the burning of coal (US NAS/NRC, 1976). This level of industrial emission of selenium may be compared with a projected annual use of selenium as a feed additive for chickens and swine in the "low-selenium" areas of the USA of 6000 kg (US FDA, 1974). But there is no evidence that the selenium deposited in rainwater has any influence on the concentration of selenium in forage plants grown in Denmark or in the heavily industrialized northeastern USA (Kubota et al., 1975). Furthermore, selenium deficiency has been observed in farm livestock grazing near coal-burning electric power stations (Anonymous, 1975). This suggests that airborne selenium exists in the form of inert elemental selenium, insoluble selenide salts, or as selenium dioxide, which would be tightly bound by acidic, iron-bearing soils. In any case, the selenium would not be available for uptake by plants; thus, there would appear to be little prospect of adding appreciable selenium to the food supply via atmospheric contamination.

(b) Human activities that decrease selenium levels in the food chain

The metabolic antagonism between sulfur and selenium (Levander, 1976a) led early workers to suggest that borderline selenium deficiency might be exacerbated by sulfate fertilization (Schubert et al., 1961). More recent research has demonstrated that sulfate fertilization can result in decreased selenium concentrations in forage plants (Pratley & McFarlane, 1974; Westermann & Robbins, 1974). In many cases, this decrease can be largely explained by a dilution effect caused by a growth response of the plant to the sulfate fertilization, but decreases in selenium concentration independent of crop yield have also been reported. Such decreases could be the result of competitive interference by sulfate with the uptake of selenate by plants. The use of phosphate fertilizers in parts of Australia and New Zealand has resulted in an apparent increase in the incidence of selenium deficiency in livestock (Judson & Obst, 1975), but others have reported increases in the selenium content of forages after fertilization with phosphorus (Robbins & Carter, 1970; Carter et al.,

1972). These conflicting results may be due to differences in the selenium content of the phosphatic rock from which the fertilizers were prepared.

Another possible way in which man's activities might decrease selenium in the food chain or render it less available, is through heavy metal pollution. For example, silver has been demonstrated to antagonize selenium in a wide variety of situations (Diplock, 1976). In an area where there has been a large amount of silver mining activity, muscular dystrophy of the type usually associated with selenium deficiency has been reported in calves fed a ration based on milk powder, even though the selenium content of the milk seemed to be adequate. Apparently, the milk powder was derived from cows that grazed land containing high levels of silver residues from the old mining industry. Further research is needed to establish fully the role of silver in potentiating these field cases of apparent selenium deficiency. Because of the many interactions of selenium with other environmental pollutants such as mercury, cadmium, or arsenic (section 7), it is possible that other cases of heavy metal-induced selenium deficiency will be discovered.

The influence of cooking on the selenium content of foods was determined because of the well-known instability and volatility of many selenium compounds. Certain vegetables that normally contain high levels of selenium such as asparagus or mushrooms, lost up to 40% of their selenium content as a result of boiling (Higgs et al., 1972). Majstruk & Suchkov (1978) reported that up to 50% of the original selenium was lost from vegetables and dairy products during cooking; the addition of salt and acid pH or extended cooking, particularly promoted such losses. But, other typical cooking procedures such as boiling cereals, baking poultry or fish, or broiling meats had little effect on selenium levels in the food (Higgs et al., 1972). Similar results were observed by others for baking fish or broiling meats, though some decrease in selenium concentration was caused by frying organ meats or fish (Ganapathy et al., 1975). Thompson et al. (1975) determined the selenium contents of cooked food composites and found a rough agreement with the calculated values based on unprepared foods from which they were derived. On the basis of all these studies, it can be concluded that usual cooking procedures cause little decrease in the selenium content of most foods.

4.1.2   Chemical forms of selenium in food

There is little information on the chemical forms of selenium that occur in human food (Levander, 1986). Cappon & Smith (1982) reported that canned tuna fish contained variable percentages of hexavalent selenium, ranging from 7.6 to 44.8%, which was independent of the total selenium content. The other forms of selenium in the tuna were di- and quadrivalent selenium. On an average percentage basis, hexavalent selenium was more easily extractable by water than di- or quadrivalent selenium. In recently canned samples, an average of 55.6% of the total selenium content was water-extractable. However, for the older samples, the corresponding average extractable level was 48.2%. The results suggest that sample storage may influence the chemical form of selenium in canned tuna. Olson et al. (1970) found that, in a certain portion of high selenium wheat (the pronase hydrolysate of gluten), about half of the selenium was in the form of selenomethionine. About 15% of the selenium in water extracts of seleniferous cabbage leaves was in the form of selenomethionine, but appreciable amounts of other selenium compounds were also present (Hamilton, 1975). The chemical forms of selenium in food are likely to affect the bioavailability of selenium (section 7.2.2 and 8.2.3).

4.2 Drinking-Water

Analysis of samples taken from various public water-supply systems in the USA showed that less than 0.5% contained selenium levels that exceeded the US PHS limit of 10 µg/litre (Taylor, 1963; Lakin & Davidson, 1967; McCabe et al., 1970). Samples from 1280 central water sources providing water for 6.5 million Bulgarians contained less than 2 µg selenium/litre (Gitsova, 1973). Tap water from Stockholm, Sweden contained only 0.06 µg/litre (Lindberg, 1968) and tap and mineral waters from Stuttgart, Federal Republic of Germany contained 1.6 and 5.3 µg/litre, respectively (Oelschlager & Menke, 1969). Surface water sources in different sub-regions of the Cernovici region of the Ukrainian SSR contained selenium levels ranging from 0.09 ± 0.01 to 3.00 ± 0.40 µg/litre; ground-water levels ranged from 0.07 ± 0.0 to 4.00 ± 0.85 µg/litre (Suchkov & Kacap, 1971). A few tenths to several µg per litre were reported in non-seleniferous regions of the USSR, the maximum reported value being 5.1 µg/litre. In Argentina, the selenium content of 22 surface waters varied from < 2 to 19 µg/litre with a median value of 3 µg/litre (WHO, 1984).

Elevated levels of selenium (> 100 µg/litre) can be found in seeps, springs, and shallow wells, though waters from deep wells contain only a few µg/litre (US NAS/NRC, 1976). Highly variable amounts of selenium were reported in wells from seleniferous areas in the USA ranging from non-detectable levels to 330 µg/litre (Smith & Westfall, 1937). Some well waters contained enough selenium to be considered as poisonous for man or livestock and loss of hair and nails in children was attributed to selenosis, but the evidence for this was not considered convincing (US NAS/NRC, 1976).

The average selenium content of 11 samples of drinking-water taken from a high-selenium area in China with a history of disease reported as chronic selenosis was 54 µg/litre (Yang et al., 1983). Four of these samples were surface water from a village with previous heavy intoxication and these averaged 139 (117 - 159) µg/litre. Such water would contribute substantial amounts of selenium, but would still comprise a small fraction of the total selenium intake in this area, since the contribution from food was estimated to be about 5000 µg/day (section 5.1.1.1). The other 7 samples of drinking-water originated from several sources and averaged only 5 µg/litre.

Rosenfeld & Beath (1964) concluded that selenium does not occur in water in sufficient amounts to produce selenium toxicity in man or animals, except in isolated cases. This was confirmed and expanded in the statement by US NAS/NRC (1976, 1980) that waters are rarely a significant source of selenium from either a nutritional or a toxicity point of view. The low concentrations of selenium in drinking-water are probably the result of several mechanisms that act to decrease the selenium content of waters (section 2.1).

4.3 Air

Selenium levels in the air breathed by the general population are probably well below 10 ng/m³, on average (US NAS/NRC, 1976). Hashimoto & Winchester (1967) found 0.3 - 1.6 ng airborne selenium/m³ in an urbanized area. Levels of 3.6 - 9.7 ng selenium/m³ were detected by Pillay et al. (1971), who showed that over half of the selenium was not retained on filters able to trap particulates greater than 0.1 µg in diameter. Zoller & Reamer (1976) concluded that atmospheric levels of selenium in most urban regions vary from 0.1 to 10 ng/m³. Selenium levels in air around industries that use selenium can be higher (section 3.2.2), perhaps of the order of a few µg/m³. Work-place air in selenium industries apparently contained mg levels of selenium/m³ in the past, but more recent measurements have indicated lower levels (section 5.2).

5. HUMAN EXPOSURE

5.1 Estimate of General Population Exposure

As discussed in section 4, selenium levels in air are low in areas without selenium-emitting industries and are not likely to exceed 10 ng/m³. Assuming that a person respires 20 m³ air daily, the contribution to daily selenium intake via this route would be 0.2 µg or less. Since this intake is much lower than that through food (see below), airborne selenium makes a negligible contribution to the average daily selenium intake of the general population. However, levels of selenium may be somewhat higher in the atmosphere around some industrial plants using selenium (section 3.2.2). If it is assumed that 3 µg selenium/m³ is found in the air near these industries, then exposure to selenium via air would be 60 µg/day.

Other possible exposures of the general population to selenium via the air include contributions from house dust and tobacco smoke. Lakin & Davidson (1967) reported that house or office dust could contain up to 10 mg selenium/kg. Although no data are available to provide an estimate of human exposure to selenium from this source, the possibility of such exposure should be considered because of the increased interest in the quality of indoor air. Olsen & Frost (1970) found an average of 0.08 mg selenium/kg (range 0.03 - 0.13 mg/kg) in a variety of cigarette tobaccos. If it is assumed that a cigarette contains 1 g tobacco and that all the selenium in tobacco is volatilized and inhaled during smoking, it can be calculated that a person smoking one pack of 20 cigarettes per day would inhale an average of 1.6 µg from this source.

Most drinking-water supplies contain only small quantities of selenium, except possibly for those in certain seleniferous areas. As discussed in section 4.2, most public water supplies contain much less than 10 µg selenium/litre. Assuming that a person drinks 2 litres of water daily, the intake via this route would be only a few µg. While this is considerably higher than the intake via air, in most areas of the world, this is still a small fraction of the daily intake from food.

5.1.1 Food

5.1.1.1 Geographical variation

Both the amount and the bioavailability of dietary selenium are determinants of its biological effects. Bioavailability is be considered in sections 7.2.2 and

8.2.3. The estimated daily intake of selenium from dietary sources by adult human beings varies considerably in different parts of the world. The greatest extremes in intake have been found in China (Table 7) where both selenium toxicity and deficiency have been reported (section 8.1.1.1, 8.2.2). An average selenium intake of 4990 µg/day was estimated in a high-selenium area of China with a history of intoxication reported as chronic selenosis, whereas an average intake of only 11 µg/day was estimated in an area of China where Keshan disease, a cardiomyopathy related to low selenium status, was reported. Intermediate intakes of selenium were found in areas of China that were not affected with either selenosis or Keshan disease. It should be pointed out that the dietary-selenium intake in the high-selenium area with a history of intoxication reported as selenosis did not overlap with that in the high-selenium area reported to be without selenosis. Likewise, the dietary-selenium intake in the low-selenium area with Keshan disease did not overlap with that in the moderate-selenium area (Beijing).

Less extreme, but still widely diverse intakes of selenium have been observed in other parts of the world (Table 8). The selenium intake calculated in New Zealand, a country with low levels of selenium in the soils in certain areas, averaged 30 µg/day (Thomson & Robinson, 1980; Watkinson, 1981). Similar low intakes have been reported by other research workers in New Zealand. For example, Stewart et al. (1978) showed that the mean dietary-selenium intake of 4 New Zealand women consuming normal diets ad libitum was 24.2 µg/day, and Griffiths (1973) reported that the daily intake of selenium by 13 young New Zealand women varied from 6 to 70 µg. Daily intakes of 30 µg or more in the latter group were associated with the inclusion of liver, kidney, or fish in the diet. A low average dietary-selenium intake (30 µg/day) has also been reported in Finland, another country known to have soils low in selenium (Varo & Koivistoinen, 1980). Low levels of dietary selenium may also occur in Italy (Rossi et al., 1976) and Egypt (Waslien, 1976). The average daily intake of selenium in the United Kingdom was estimated to be 60 µg (Thorn et al., 1978). In one study, the dietary-selenium intake in Japan was estimated to be 88 µg/day (Sakurai & Tsuchiya, 1975). A higher estimate in another study was arrived at largely because higher selenium contents of food staples were used in the calculations (Yasumoto et al., 1976). The estimated dietary intake of selenium in North America ranged from 98 to 224 µg/day. A "Market basket" survey carried out in the USA from 1974 to 1982 indicated an average overall selenium intake of 108 µg/day (Pennington et al., 1984). An earlier survey had revealed regional differences in intake, with people living in western parts of the

Table 7. Daily selenium intake of residents living in high-, medium-, and low-selenium areas of China[a]

Place	Number of subjects	Daily selenium intake (µg)			Se intake from staple cereals as % of total daily intake
		minimum	maximum	average	
High-selenium area with a history of intoxication reported as chronic selenosis	6	3200	6690	4990	28-70%
High-selenium area reported as without selenosis	3	240	1510	750	25-45%
Moderate-selenium area (Beijing)	8	42	232	116	various sources
Low-selenium area with Keshan disease	13	3	22	11	mainly from cereals

[a] Adapted from: Yang et al. (1983).

Table 8. Estimated human dietary intake of selenium (µg/day)[a]

Food	New Zealand Dunedin[b]	New Zealand Hamilton[c]	Finland 1975[d]	Finland 1979[d]	United Kingdom[e]	Japan[f]	Canada[g]	USA[h]	USA South Dakota[i]	Venezuela[j]
Plant										
vegetables, fruits, and sugars	1	2	1	1	3	6	1-9	5	10	15
cereals	4	3	3	25	30	24	62-113	45	57	88
Animal										
dairy products, eggs	11	11	7	13	5	2	5-28	13	48	70
meat, fish	12	16	19	19	22	56	25-90	69	101	153
Total	28	32	30	50-60	60	88	98-224	132	216	326

a From: Robinson & Thomson (1983).
b From: Thomson & Robinson (1980).
c From: Watkinson (1981).
d From: Varo & Koivistoinen (1981).
e From: Thorn et al. (1978).
f From: Sakurai & Tsuchiya (1975).
g From: Thompson et al. (1975).
h From: Watkinson (1974) & Levander (1976b).
i From: US NAS/NRC (1980).
j From: Levander (1976b).

country consuming 1.3 times as much selenium as people in the northeastern part of the country (US FDA, 1975). Duplicate plate analyses of self-selected diets in Maryland, USA gave a mean intake of 81 ± 41 µg/day (Welsh et al., 1981). The daily intake in Venezuela was estimated to be 326 µg, but this figure requires cautious interpretation because north American food consumption patterns were used in its derivation (Mondragon & Jaffe, 1976).

5.1.1.2 Food habits (consumption patterns)

Because of the wide variations in the selenium content of various food groups in different countries, it is possible that certain food habits or preferences could play an important role in determining whether a person is at risk with regard to a deficient or excessive intake of selenium. Persons with a preference for foods such as fish may consume high levels of selenium in their diet. Sakurai & Tsuchiya (1975) estimated that Japanese eating large amounts of seafish may ingest as much as 500 µg of selenium daily. People residing in seleniferous agricultural zones may consume large amounts of selenium if locally-produced foods constitute the bulk of their diet. In Venezuela, for example, corn and rice samples from high selenium areas contained as much as 14 and 18 mg selenium/kg, respectively (Jaffe, 1976), and values as high as 180 mg/kg have been reported in wheat from Colombia (Ancizar-Sordo, 1947). If it is assumed that some segments of the population living in such seleniferous regions consume, under certain conditions, about half of their diet in the form of rice or corn with a similar selenium content to that described above, the daily intake of selenium from dietary sources could exceed 7000 µg. However, it should be pointed out that only a few samples from the high-selenium areas showed these extreme high levels. Obviously, vegetarian diets could vary tremendously in selenium content. Ganapathy & Dhanda (1976) showed that a vegetarian diet in the USA supplied 84 µg selenium/day.

If a particular staple food constitutes a large fraction of the diet of a population, then the selenium content of that food will have a great influence on the overall dietary-selenium intake. For example, one estimated daily dietary intake of selenium in Japan (Sakurai & Tsuchiya, 1975) of 88 µg, was based on selenium levels in rice and soybeans of 0.05 and 0.02 mg/kg, respectively, whereas another estimated intake, 208 µg, was based on rice and soybean selenium levels of 0.220 and 0.234 mg/kg, respectively (Yasumoto et al., 1976). Low levels of selenium in rice have been reported in Bangladesh, and it would be particularly interesting to study the population there, as they depend on rice as a staple food, but they have a recognized poor dietary vitamin E status (Bieri & Ahmad, 1976).

The importance of the selenium content of a particular staple in the food supply of persons dependent on a monotonous diet consisting of locally-produced foods was recently pointed out in connection with the occurrence of Keshan disease (section 8.2.2) in China (Yu, 1982). In the years in which Keshan disease was endemic in the Fuyu county of Heilongjiang province, about 90% of the diet consisted of maize, whereas, in the years in which Keshan disease was not endemic, only about 30% of the diet was maize with much of the rest coming from millet and wheat. Thus, it seems possible that over-reliance on this single staple food during certain years helped to precipitate Keshan disease in this area.

Recently, there has been considerable interest in some countries in the use of textured soy protein products as substitutes for meat products. Since meat products are such good dietary sources of selenium, the possible impact of this trend on the selenium intake of human beings was examined. Some soy protein-based meat analogues contained as much selenium as meat products, but others did not (Ferretti & Levander, 1976). These results show that meat analogues made from soy or other plant proteins are not consistent sources of selenium, since these plant-based products will vary in selenium content depending on where the plant is grown. In contrast, meat products are more reliable dietary sources of selenium, because they contain a minimum quantity compatible with animal life.

5.1.1.3 Elderly people

Thomson et al. (1977a) studied a group of 48 elderly people (mean age of 78 years) in New Zealand and found that their blood-selenium levels and red cell glutathione peroxidase (EC 1.11.1.9) activity were lower than those in 12 young adult controls. It was not possible for the authors to conclude from the data whether the depressed blood-selenium levels and enzyme activity were due to decreased selenium intake or were a general part of the aging process. Abdulla et al. (1979) analysed 7-day pooled dietary composites of 20 pensioners in Sweden and found that the average daily selenium intake was 30.8 µg with a range of 8.7 - 96.3 µg/day.

5.1.1.4 Infants and children

The selenium intake of infants is of interest because of their rapid growth rate and their heavy reliance on milk, a food that has a highly variable selenium content, depending on its geographical origin (Table 9). Millar & Sheppard (1972)

Table 9. Estimated infant daily intake of selenium from dietary sources in North America (6-month-old, 6.5-kg child)[a]

Food	Daily consumption (g)	Selenium content (mg/kg)	Selenium intake (μg/day)
Milk	824	0.013	11
Orange juice	122	0.014	2
Dry mixed cereal	10	0.540	5
Egg yolk	17	0.437	7
Strained meat	28	0.097	3
Strained fruit	57	0.002	-
Strained vegetable	57	0.003	-
Total selenium intake			28

[a] Adapted from: Levander (1976b).

assumed that human and cow's milk in New Zealand contained 0.010 and 0.005 mg selenium/litre, respectively, and calculated that the selenium intake of infants for the first month of life would be 5.0 or 2.1 μg/day, depending on whether human or cow's milk was fed. Williams (1983) measured the selenium content of mature mother's milk from 10 New Zealand women and calculated an average daily selenium intake of 5 - 6 μg for an infant consuming 700 ml milk. If the value for the selenium content of Venezuelan cow's milk shown in Table 9 is typical, it can be calculated that the selenium intake of infants in that country would be about 58 μg/day during the first month of life. The most extreme values for the selenium content of human milk are from low- and high-selenium areas of China (Yang et al., 1986) (section 4.1.1.1). For infants consuming 750 ml milk daily, these milks would provide 2 μg and 212 μg selenium, respectively.

It was shown that milk continues to furnish a sizeable portion of an infant's estimated daily selenium intake of 28 μg, in the USA, after the child starts to eat solid foods (Table 9). The estimated daily intake of selenium by 3-month-old infants in the Federal Republic of Germany was 11 μg (Lombeck et al., 1975) and a "market basket" survey indicated that the intake of 6-month-old infants in the USA was about 12 μg/day (Harland et al., 1978). On the basis of a caloric intake of 700 Kcal/day, infants in the USA consuming formula diets based on milk, soy protein, casein, or meat, ingested 8.5, 9.5, 12.6, or 31.5 µg/day, respectively (Zabel et al., 1978).

As infants grow up in New Zealand, they continue to ingest relatively low levels of selenium, since the estimated daily intake over the age span of 6 months to 5 years was 5 - 8 µg (McKenzie et al., 1978).

5.1.1.5 Special medical diets

Infants and children suffering from metabolic diseases such as phenylketonuria (PKU) or maple syrup urine disease (MSUD) are fed synthetic diets that contain little selenium for the first 10 or 12 years of their lives. In New Zealand, the mean daily selenium intake of 12 subjects consuming such synthetic diets was 5 ± 2 µg with little difference between the older children and the infants (McKenzie et al., 1978). The selenium content of diets of 3-month-old infants with PKU and MSUD in the Federal Republic of Germany amounted to 4.7 and 5.6 µg/day, respectively (Lombeck et al., 1975).

A wide variation in the selenium contents of special-purpose medical diets for adults is related to the source of protein in the diet (Zabel et al., 1978). On the basis of an intake of 2000 Kcal/day, adults consuming food supplements or tube-feeding formulae would ingest 29, 47, or 98 µg selenium/day, depending on whether the diet contained soy protein and casein, milk protein and casein, or blended foods. At a similar caloric intake, chemically-defined diets based on egg albumen provided 225 µg/day, whereas diets based on casein hydrolysate or amino acid mixtures provided only 28 and 1.5 µg/day, respectively. Total parenteral nutrition solutions based on casein hydrolysate and amino acid mixtures also furnished low amounts of selenium of only 32 and < 1 µg/day, respectively. When levels of selenium supplied by special therapeutic diets were studied in the Chernovitsi region of the Ukrainian SSR, selenium intakes were found to vary between 99 ± 6 and 121 ± 8 µg/day (Majstruk & Suchkov, 1978).

5.2 Occupational Exposure

5.2.1 Levels in the work-place air

The main pathway of human occupational exposure to selenium is through the air. Rarely, and under special circumstances, direct contact may be of importance, i.e., when cutaneous absorption might be facilitated by the local irritation and skin damage caused by vesicant selenium compounds.

Various mechanical processes connected with the mining of seleniferous ores or the grinding of selenium compounds can contribute selenium-containing dusts to the atmosphere. In other industrial activities, the amount of selenium released into the air depends on the temperature to which it is heated and on the area available for sublimation and/or vaporization. It is well established that heating amorphous selenium below its melting point results in its sublimation. At temperatures of 170 - 180 °C, traces of selenium can be detected in the air and, at temperatures of 230 - 240 °C, selenium dioxide is released. Heating of amorphous selenium and selenium dioxide resulted in the release of substantial quantities of selenium into the air, e.g., from a surface of 10 cm², 2.6 - 3.75 mg was released after 20 min at 230 - 240 °C and 23 - 31.6 mg was released after 20 min at 350 - 360 °C (Izraelson et al., 1973).

Little quantitative information is available in the literature on actual levels of human exposure to selenium in industry. The levels of selenium in the work-place depend not only on the nature of the process involved but also on the control technology in any given industry. Thus, it can be expected that analyses carried out several decades ago yielded higher values than analyses carried out more recently. In 1948, Filatova reported air levels of selenium and selenium dioxide of 20.6 - 24.8 mg/m³ and 0.46 - 0.78 mg/m³, respectively, in working zones of selenium-heating processes in a selenium rectifier plant, whereas levels of 0.55 - 1.1 mg/m³ and up to 0.11 mg/m³, respectively, were associated with the process of disc smearing (Filatova, cited in Izraelson et al., 1973). In the production of selenium-containing photoelements by the vacuum application of a photosensitive layer, selenium and selenium dioxide levels did not exceed 2.0 mg/m³ and 0.1 mg/m³, respectively (Sverdlina & Maslennikova, 1961). Selenium grinding, sulfuric acid treatment, and cathode alloy coating resulted in air-selenium levels ranging from 0.133 to 2.0 mg/m³ (Izraelson et al., 1973). A highly-dispersed aerosol selenium condensate with a particle size of 0.5 - 2 µm was produced at a level of 0.88 - 6.13 mg/m³, as a result of the manufacture of selenium-containing steels (Ershov, 1969). In the production of rare metals, the air in industrial premises can contain as much as 0.4 - 750 mg metallic dust/m³, which, in addition to other elements, contains from 0.8 - 15.7% selenium (Burchanov, 1972). During the 1950s, Glover (1967) carried out an extensive survey in which air-selenium levels were determined in areas of a selenium rectifier plant in which the workers had been found to have high urinary-selenium values. In general, the highest air levels of 1.5 - 5.2 mg selenium/m³ were found in relation to the grinding process. With the exception of certain "special processes", air levels in the plant due to all the other manufacturing activities were less than 0.5 mg/m³. In one case, a level of 21.3 mg/m³ was observed, but no additional details were given. In this rectifier plant, a process, in which the elemental selenium was heated, was used for preparing the selenium mixture that was later applied to the metal plate. For this reason, the selenium levels in the air are likely to have been higher than in other factories where the selenium was deposited from aqueous solution. In another selenium rectifier plant, air levels of between 7 and 50 µg selenium/m³ were reported (Kinnigkeit, 1962). However, in this case, no correlation was observed between the air levels in the work-place and the selenium levels in the blood of the workers. Kinnigkeit felt that the air-selenium levels were not high enough to account for the high urinary-selenium levels in the workers. He suggested that the discrepancies might be due to incomplete collection of selenium during air sampling, transient peaks of selenium exposure that were not detected by short-term sampling, or contamination of hands, body, and clothing with selenium, which was then transferred to the mouth as a result of eating or smoking. Another possible explanation is that measurements of the levels in the work-room air did not accurately reflect levels of selenium in the actual breathing zone of the workers.

Low air levels of selenium have been observed in factories involved with the use of selenium in certain photocopying devices. In this type of manufacturing, "clean room" techniques are used, because strict dust control is needed to produce the quality of photoreceptor surfaces required. Cannella (1976) reported air levels of selenium ranging from 6 to 91 µg/m³ in various locations in a selenium alloy plant that made such photoreceptor products.

5.2.2 Biological monitoring

Although selenium levels in the work-place air have been determined in several instances, these measurements cannot be used to draw any conclusions about the exposure history of individual workers. Glover (1967), who estimated occupational exposure to selenium indirectly by determining selenium levels in the urine of workers, acknowledged the limitations associated with monitoring based on grab samples of urine. The Working Group also recognized the inadequacy of urinary-selenium concentrations as a monitoring tool. However, it appears that the use of a more rigorous programme (e.g., total 24-h urine collections) would have resulted in losing the cooperation of the workers involved. A tentative maximum allowable concentration of selenium in urine of 0.1 mg/litre was proposed by Glover (1967). The urine of the workers was analysed at 3-month intervals, and any employee whose urinary-

5

selenium value exceeded the limit was transferred to a process not involving selenium. The urinary-selenium levels of these individuals were tested weekly.

Two primary factors were responsible for setting the maximum allowable concentration for selenium in urine at 0.1 mg/litre (Glover, 1970). First, a limit lower than this would have resulted in a certain number of "false positives", since selenium levels in urine are often this high in people not industrially exposed to the element. Second, it was found that values below 0.1 mg/litre were invariably accompanied by air levels in the work-place of less than 0.1 mg/m^3, which was below the threshold limit value of 0.2 mg/m^3 for elemental selenium and its common inorganic compounds (ACGIH, 1971). It must be emphasized that urinary selenium cannot be used to monitor exposure to such dangerous selenium compounds as hydrogen selenide, selenium oxychloride, or certain organic selenium compounds because severe damage to the lungs or skin would have occurred by the time that the urinary-selenium values were raised (Glover, 1976).

6. METABOLISM OF SELENIUM

The metabolism of selenium has been most completely studied in animals and data from human investigations are limited. Thus, in the following sections, pertinent information from selected animal studies is presented followed by available information from controlled human studies.

6.1 Absorption

Since food is the primary environmental medium through which man and animals are exposed to selenium, most data concerning selenium absorption deal with the gastrointestinal pathway. Much less is known about the absorption of selenium through the lungs or skin.

6.1.1 Gastrointestinal absorption

6.1.1.1 Animal studies

The intestinal absorption of soluble selenium compounds by rats is highly efficient. It has been shown that these animals absorbed 92, 91, and 81% of doses of selenite, selenomethionine, and selenocystine, respectively (Thomson & Stewart, 1973; Thomson et al., 1975a). In a study on rats by Whanger et al. (1976a), the greatest absorption of selenite or selenomethionine occurred from the duodenum, with slightly less absorption from the jejunum or ileum. Virtually no absorption occurred from the stomach. The absorption of selenite by rats does not appear to be under homeostatic control, since 95% or more of the dose was absorbed, regardless of whether the animals were fed a selenium-deficient diet or a diet containing mildly toxic levels of selenium (Brown et al., 1972). Wright & Bell (1966) found that selenite was absorbed from the gastrointestinal tract to a greater extent by monogastric animals (swine) than by ruminant animals (sheep). The decreased absorption by sheep may have been due to reduction of the administered selenite to insoluble or unavailable forms by rumen microorganisms.

Only limited information is available regarding the absorption by animals of selenium occurring naturally in foods. Kidney tissue and fish muscle were chosen for model studies, because these foods are relatively rich sources of dietary selenium for human beings. The levels of intestinal absorption by rats of radio-selenium administered in homogenates of rabbit kidney or fish muscle injected with selenite, were 87 and 64%, respectively (Thomson et al., 1975b; Richold et al., 1977). The low absorbability of "fish

selenium" agrees with other reports showing the poor bioavailability to animals of selenium in certain fish products
(section 5).

Little is known concerning the physiological processes
governing the absorption of even simple selenium compounds,
though McConnell & Cho (1965) showed that selenomethionine was
transported against a concentration gradient, whereas selenite
and selenocystine were not.

6.1.1.2 Human studies

Studies carried out in New Zealand on 3 young female
volunteers indicated that the intestinal absorption of oral
doses of radioactive selenite, containing not more than 10
µg selenium, was 70, 64, and 44% of the dose, respectively
(Thomson & Stewart, 1974). In a study conducted 2 years
later, in which 2 of the 4 female volunteers were women who
had participated in the earlier study with selenite, the
intestinal absorption of oral doses of ^{75}Se-seleno-
methionine, containing less than 2 µg selenium, averaged
about 96% of the dose (Griffiths et al., 1976). When a larger
dose of 1 mg selenium was given orally in solution, a similar
difference in absorption was obtained, i.e., 97% for selenomethionine (one subject) (Thomson et al., 1978) and about 60%
for sodium selenite (13 subjects) (Robinson et al., 1985;
Thomson & Robinson, 1986). This value for absorption of
selenite-selenium is much lower than the 92% absorption
reported for "selenite" in the earlier paper of Thomson
(1974), which is now known to have been selenate-selenium and
not selenite-selenium. Later studies on 10 volunteers also
showed absorption of 94% for 1 mg selenate-selenium in
solution (Thomson & Robinson, 1986). Thus, selenate-selenium
is similar to selenomethionine-selenium in that it is better
absorbed than selenite-selenium.

Correction of selenate-selenium for selenite selenium in
the paper of Thomson (1974) also removes the "surprising
difference" reported between absorption of selenite given in
solid form (60% for 4 female volunteers) and in solution (60%
and not 92%). A female volunteer given a 1 mg dose of
selenium as selenite in solution daily for 5 consecutive days
absorbed 59% of the total 5 mg dose (Thomson et al., 1978).
It would appear that human beings have no homeostatic control
to limit their absorption of large single doses of these
selenium compounds (Barbezat et al., 1984).

Robinson et al. (1978b) measured the intestinal absorption
of selenium by a New Zealand female volunteer receiving a
daily supplement of 100 µg selenium as selenomethionine, a
male volunteer supplemented daily with 100 µg selenium as
sodium selenite, and a second female supplemented daily with

65 µg selenium, as it occurs naturally in mackerel (Scomber japonieus). The selenomethionine and selenite were given in solution, whereas the mackerel was given as canned fish. During a 4-week supplementation period, 75% of the selenium given as selenomethionine was absorbed compared with only 48% of the selenium administered as selenite. The absorption of the "fish selenium" was 66%. Details of the diets consumed by the 3 subjects were not provided, but the subjects supplemented with selenomethionine or fish selenium did not consume liver or kidney throughout the study. However, it was specified only that the subject supplemented with selenite did not eat liver, kidney, or fish on the days that weekly urine collections were made or on the day prior to the collection. Also, the subject receiving selenomethionine ate fish only occasionally throughout the study. The subject supplemented with the fish selenium ate fish other than mackerel on only one occasion. Stewart et al. (1978) found that the average intestinal absorption of naturally-occurring selenium in foods was 55% in 4 young New Zealand women freely consuming normal diets. However, allowing for the endogenous faecal excretion (half of total faecal output) the average true absorption of naturally occurring selenium became 79%.

6.1.2   Absorption by inhalation

Certain volatile selenium compounds, such as hydrogen selenide, are known to be toxic when inhaled (Dudley & Miller, 1941). Lipinskij (1962) observed increased selenium levels in the liver and kidneys after respiratory exposure of rabbits to elemental selenium and selenium dioxide. More recently, Weissman et al. (1979, 1983) studied the distribution and retention of inhaled ^{75}Se-labelled selenious acid and elementary selenium aerosols. They used 63 Beagle dogs, which were between 3 and 4 years old at the time of exposure. All inhalation exposures were through the nose only. Aerosol particles of both forms were small with median aerodynamic diameters of 0.5 and 0.7 µm and geometric standard deviations of 2.4 and 1.5 for selenious acid and elemental selenium, respectively. The urine and faeces of 4 dogs were collected daily, for 32 days, and then at regular time intervals until sacrifice. Selenium from exhaled breath was collected from 3 dogs for several 4-h periods, between 2 and 10 days after exposure. Whole-body retention of ^{75}Se-selenium was measured immediately after exposure and at regular intervals up to 320 days after exposure. Two dogs exposed to each aerosol were sacrificed at various time intervals for over half a year. The initial ^{75}Se-selenium body burden (IBB) in dogs that inhaled ^{75}Se-selenium as selenious acid was 40 ± 17 µg selenium/kg body weight and 22 ± 9 µg selenium/kg body weight in dogs inhaling elementary selenium. For both forms of selenium, the retention data of ^{75}Se-selenium could be expressed by a 3-component negative exponential equation:

Selenious acid:

$$\%IBB = 66e^{-1.5t} + 17e^{-0.089t} + 17e^{-0.023t}$$

Elemental selenium:

$$\%IBB = 66e^{-0.090t} + 21e^{-0.059t} + 13e^{-0.018t}$$

The corresponding half time values were 0.46, 7.8, and 30 days and 0.77, 12, and 38 days for selenious acid or elemental selenium exposure, respectively. Studies on organ distribution showed that, 2 h after exposure to selenious acid, only about 5.3% of the IBB was retained in the lung compared with 26% of the IBB of elemental selenium. For the first 32 days following exposure, the urine accounted for 79 and 66% of the ^{75}Se-selenium excreted by dogs exposed to selenious acid and elemental selenium aerosols, respectively. Respiratory excretion of ^{75}Se-selenium was negligible. The organ distribution of ^{75}Se-selenium as a percentage of the sacrifice body burden is given in Table 10.

Comparison of the absorption of these two forms of selenium through inhalation was also studied in the rat (Medinsky et al., 1981). The rate of transfer of selenium into the blood was slower when the elemental form was administered. However, once absorbed, both chemical forms behaved identically.

6.1.3   Absorption through the skin

Apart from the report of Dutkiewicz et al. (1972), who found that 10% of a 0.1 mol solution of sodium selenite applied to rat skin was absorbed in 1 h, there are no quantitative data on the dermal absorption of water-soluble selenium compounds. However, Dudley (1938) showed that selenium oxychloride was absorbed through the skin of rabbits. Selenium sulfide, an insoluble selenium compound used in certain anti-dandruff shampoos, is not ordinarily absorbed through the skin, although a garlicky breath odour and elevated urinary excretion of selenium were noted in a patient with open scalp lesions who used a selenium sulfide shampoo (Ransone et al., 1961).

Table 10. ^{75}Se-selenium distribution in organs of dogs 128 days
after respiratory exposure to ^{75}Se-selenious acid or elemental
^{75}Se-selenium aerosols[a]

Organ	^{75}Se-selenium organ distribution as % of ^{75}Se-selenium body burden at sacrifice	
	selenious acid treated	elemental selenium treated
Lung	1.3	6.2
Liver	11.4	7.4
Kidney	1.6	1.6
Blood	19.7	21.5
Gastrointestinal tract	3.4	3.6
Pelt	9.6	9.1

[a] Adapted from: Weissman et al. (1979).

6.2 Distribution in the Organism

6.2.1 Transport

Little is known about the transport of selenium in the
body. However, a plasma selenoprotein has been identified
which one group postulates is involved in selenium transport
(Motsenbocker & Tappel, 1982).

6.2.2 Organs

6.2.2.1 Animal studies

Absorbed selenium is rapidly distributed among the
tissues. Heinrich & Kelsey (1955) found that the liver of
mice injected subcutaneously with sodium selenite contained 17
and 30% of the dose, 15 and 60 min after injection, respectively. However, because of the great variation in the metabolic half-time of selenium in various tissues (Thomson &
Stewart, 1973), the distribution pattern observed will change
markedly with the time of sampling after injection with radio-
selenium.

As long as the diet provides nutritionally adequate levels
of selenium, the relative concentrations of selenium in the

various internal organs are quite consistant over a wide range
of selenium intakes. For example, Jones & Godwin (1962)
studied the distribution of selenium in the tissues of mice
fed alfalfa containing nutritional levels of selenium. The
concentrations of selenium in the internal organs decreased in
the following order: kidneys > liver > pancreas >> lungs
> heart > spleen > skin > brain > carcass. Smith et
al. (1937) fed toxic levels of sodium selenite to cats and
observed the following order for the concentration of selenium
in the internal organs: liver > kidneys > spleen >
pancreas > heart > lungs.

The distribution of selenium appears to be relatively
independent of the form and route of administration, since the
kidneys and adrenals contained the highest concentrations of
radio-selenium, regardless of whether the rats were given
labelled selenite or selenomethionine orally or by intravenous
injection (Thomson & Stewart, 1973). The greatest amounts of
radioactivity were in the kidneys, liver, and total muscle
mass. Similar distribution patterns were noted in rats dosed
orally with homogenates of tissues from rabbits or from fish
that had been given labelled selenium compounds (Thomson et
al., 1975b; Richold et al., 1977).

The distribution of selenium in tissues was markedly
different when rats were fed a diet deficient in selenium: the
testes, brain, thymus gland, and spleen took up the greatest
concentrations of selenium from a tracer dose of selenite
(Burk et al., 1972). A recent report confirms the ability of
the testis to retain selenium better than other tissues in
selenium deficiency (Behne & Hofer-Bosse, 1984).

The distribution of selenium in the eye was determined
fluorometrically with 2,3-diaminonaphthalene by Gusejnov et
al. (1974), who found the following levels in various parts of
the eye of a bull (mg/kg fresh weight): iris, 1.3; retinal
pigment epithelium, 0.85; retina, 0.50; cornea, 0.3; lens,
0.03; and vitreous body, 0.01. In some birds (pigeons and
crows), the level of selenium in the retina was about 10 times
higher (about 5.0 mg/kg). In studies in which ^{75}Se-labelled
sodium selenite in saline was administered intraperitoneally
to frogs, rats, and rabbits, the distribution of radio-
selenium was similar to that of the naturally-occurring
selenium measured in the bull's eye. The maximum [75]Se
activity in the eye (5 h after administration) was found in
the pigment with the vascular layer of the retina, whereas
minimum activity was found in the lens and vitreous body.
Histoautoradiography of the eyes also indicated the highest
radioactivity in the vascular and pigment layers (Abdullaev et
al., 1972).

6.2.2.2 Human studies

The distribution of selenium in human organs was examined
by Schroeder et al. (1970), who found the following descending
order for selenium concentrations in tissues: kidney > liver
> spleen > pancreas > testes > heart muscle >
intestine > lung > brain. Blotcky et al. (1976) showed
that, in autopsy samples from 106 persons, the concentrations
of selenium were 2 - 3 times higher in the kidneys than in the
liver.

Uptake of an orally-administered dose of radioselenite in
the internal organs occurred in the following relative order:
liver > kidneys > lungs > muscle = articular tissue
(Leeb et al., 1977). Other studies have shown significant
accumulation of radio-selenium by the liver and kidneys after
intravenous injections of selenite or selenomethionine
(Lathrop et al., 1972; Falk & Lindhe, 1974; Nordman, 1974).

6.2.3 Blood

Studies on experimental animals have shown a close
correlation between the level of selenium in the blood and the
level of selenium in the diet (Lindberg & Jacobsson, 1970;
Cary et al., 1973; Ermakov & Kovalskij, 1974; Oh et al.,
1976a). Analytical surveys on human beings have shown
significant variations in blood-selenium levels in different
geographical areas, presumably because of differences in
dietary-selenium intake. For example, Allaway et al. (1968)
found that selenium levels in the blood of persons living in
areas of the USA where soils and vegetation were high in
selenium tended to contain more selenium than that of persons
living in areas where soils and plants were generally low in
selenium. Thus, geographical differences in the selenium
levels in human blood are possible, even in a country where
large-scale interregional food shipments would be expected to
level out differences in the amount of selenium in the food
supply. These geographical differences in human blood-
selenium levels in the USA have been confirmed by Howe (1974),
who found a mean selenium content of 0.265 mg/litre (SD =
0.056) for 626 samples collected in and around the state of
South Dakota (a high soil-selenium region), whereas Schultz &
Leklem (1983) reported blood-selenium values in Oregon (a low
soil-selenium region) that were well below those reported for
other areas of the USA.

Analyses of blood samples from persons living in different
regions of the world indicate similar correlations between the
amount of selenium in the blood and the amount of selenium in
the food throughout the world (Table 11). For example,
blood-selenium concentrations tend to be lowest in areas of

Table 11. Comparison of selenium levels in whole-blood samples
obtained from human beings living in different parts of the world
(values for each country are listed in decreasing numerical order)

Country or region	Reported values (mg selenium/litre whole blood)	Reference
Canada		
Ontario	0.182 SD ± 0.036	Dickson & Tomlinson (1967)
The People's Republic of China		
high-selenium area with a history of intoxication reported to be chronic selenosis[a]	3.2 (1.3-7.5)	Yang et al. (1983)
high-selenium area reported to be without selenosis[a]	0.44 (0.35-0.58)	Yang et al. (1983)
moderate-selenium area (Beijing)[a]	0.095 SD ± 0.091	Yang et al. (1983)
low-selenium area without Keshan disease	0.027 SD ± 0.009	Yang et al. (1983)
low-selenium area with Keshan disease[a]	0.021 SD ± 0.010	Yang et al. (1983)
Egypt	0.068 (0.054-0.079)	Maxia et al. (1972)
Finland		
Helsinki	0.081 SD ± 0.015	Westermarck et al. (1977)
Lappeenranta	0.056 SD ± 0.017	Westermarck et al. (1977)
Guatemala	0.23 SD ± 0.05	Burk et al. (1967)
New Zealand		
Auckland, North Island	0.083 SD ± 0.013	Robinson & Thomson (1981)
Dunedin, South Island	0.059 SD ± 0.012	Rea et al. (1979)
Sweden	0.12 SD ± 0.02	Brune et al. (1966)
United Kingdom	0.32 (0.26-0.37)	Bowen & Cawse (1963)

Table 11 (contd).

Country or region	Reported values (mg selenium/litre whole blood)	Reference
USA		
Rapid City, South Dakota	0.256 SE ± 0.036	Allaway et al. (1968)
Lima, Ohio	0.157 SE ± 0.032	
USSR		
Ukrainian SSR	0.442 SE ± 0.034	Suchkov (1971)
Azerbeijan SSR	0.11 SE ± 0.007	Abdullaev (1976)
Venezuela		
Seleniferous zone Villa Bruzual	0.813	Jaffe et al. (1972a)
Caracas	0.355	

a   These blood samples were collected in areas corresponding to those shown in Table 8.

the world that are known to have low levels of selenium in the soil (Keshan-disease areas of China, New Zealand, Scandinavia) and highest in areas that are known to have high levels of selenium in the soil (Venezuela, reported selenosis areas of China). Superimposed on these geographical variations in effects are the effects of general nutritional status or effects of certain disease states. For example, children suffering from Kwashiorkor have lower blood-selenium levels than their well-nourished counterparts (Burk et al., 1967) and reductions in the selenium content of sera have been observed in patients with cancer (McConnell et al., 1975). Lower blood-selenium values were associated with lower serum-albumin values in surgical patients with and without cancer in New Zealand (Robinson et al., 1979). Akesson (1985) suggested that alterations in plasma-selenium may be secondary to changes in plasma-protein concentration. In a study of Finnish men with one or more risk factors for coronary heart disease, Miettinen et al. (1983) observed a strong association between serum-selenium and the eicosapentanoic acid contents of cholesterol ester and phospholipids. The authors pointed out that since eicosapentanoate is a fatty acid peculiar to fish such an association could reflect the intake of fish at the time of blood sampling.

Blood-selenium levels change if persons travel to countries with soils of different selenium content. For example, Griffiths & Thomson (1974) noted that the blood-selenium levels of adults from the USA declined rapidly on arrival in New Zealand, but, after 1 year, their levels were still higher than the mean value for permanent New Zealand residents (Fig. 2). Rea et al. (1979) found that plasma-selenium levels changed with changes in selenium intake more rapidly than erythrocyte-selenium levels. The latter reflected longer-term changes in selenium exposure, presumably because of the relatively long life span of erythrocytes. In a controlled depletion/repletion metabolic study (section 6.3.2.2), Levander et al. (1980) stated that the plasma-selenium levels of healthy young male volunteers in the USA dropped about 19% after 2 weeks on a low-selenium diet and then returned to their previous levels 11 days after consuming a high-selenium diet (Fig. 3).

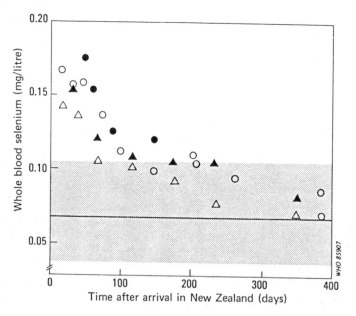

Fig. 2. Decline in whole-blood selenium concentration in adults from the USA after arrival in New Zealand. Heavy solid line represents mean concentration for New Zealand residents. Shaded area covers range of concentrations found in 170 New Zealand residents. From: Griffiths & Thomson (1974).

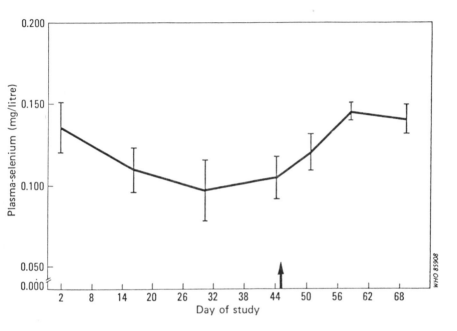

Fig. 3. Effect of experimental selenium depletion and repletion on plasma-selenium levels in young male volunteers. Arrow indicates end of depletion and start of repletion. From: Levander et al. (1980).

The Task Group was aware of one report (Kinnigkeit, 1962) that presented blood-selenium levels in workers occupationally exposed to selenium (Table 12). The analytical method used in this study was a wet chemical technique based on complexing selenium with diaminobenzidine, and the blood-selenium values obtained by this procedure from a population not occupationally exposed were below 0.5 mg/litre. These values imply that blood-selenium levels in workers, at least in the past, were much higher than those found in the general population (Table 11). Unfortunately no control values were available (Table 12) so it was not possible to compare the blood-selenium levels with those obtained using the more recent analytical technique using diaminonaphthalene instead of diaminobenzedine (section 2.2.5.1).

6.2.4   Total-body selenium content

Schroeder et al. (1970) estimated the total-body selenium content of persons in New England, USA, by multiplying the mean values of the selenium content of human tissues obtained

Table 12. Blood-selenium levels reported in workers from a
selenium rectifier plant[a]

Department	Number of workers	Selenium level in blood (mg/litre)
Vaporization	13	4.8 ± 14.8
Measurement field	13	15.8 ± 11.8
Stamping	12	8.8 ± 13.8
Electric fabrication	18	13.4 ± 12.9

[a] Geometric mean ± standard error. From: Kinnigkeit (1962).

at autopsy by standard organ weights. In this way, a total-body selenium content of 14.6 mg (range, 13.0 - 20.3 mg) was calculated for 91.7% of the body. Stewart et al. (1978) calculated the total-body selenium content of 4 New Zealand women by 3 different techniques: using the specific activity of urinary selenium and retained whole-body radio-selenium; using plasma-selenium and the occupancy of radio-selenium in whole body and plasma; and using absorbed food-selenium and the occupancy of absorbed radio-selenium in the whole body. Depending on whether labelled selenomethionine or selenite was used in the estimation, the total-body selenium content was found to be either 6.1 mg (range, 4.1 - 10.0 mg) or 3.0 mg (range, 2.3 - 5.0 mg), respectively, which was less than half that of North Americans. This was consistent with the selenium contents of individual tissues (liver, muscle, heart) (Money, 1978; Casey et al., 1982) of New Zealand subjects, which also contained less than half that reported in North Americans.

6.3 Excretion in Urine, Faeces, and Expired Air

6.3.1 Animal studies

Studies on rats have shown that the urinary pathway is the dominant route for selenium excretion, as long as the dietary selenium exceeds a certain critical threshold level. For selenium as selenite, this threshold lies between 0.054 and 0.084 mg/kg (Burk et al., 1973). As the dietary level of selenium increased from 0.004 to 1.000 mg/kg, the cumulative 10-day urinary excretion of a tracer dose of radio-selenite increased from 6 to 67% (Burk et al., 1972). In contrast, the faecal excretion of selenium remained constant at about 10% of the dose over this range of dietary-selenium intake. Significant differences in the urinary excretion of selenium were demonstrated after dosing rats orally with various forms of selenium (Richold et al., 1977). For example, the cumulative levels of selenium excreted in the urine, one week after dosing with selenite, selenocystine, selenomethionine, "rabbit kidney" selenium, and "fish muscle" selenium, were 14, 14, 5, 7, and 6% of the absorbed dose, respectively. In many of these studies, the endogenous faecal excretion of selenium approached or exceeded urinary excretion. These decreased urinary/faecal excretion ratios may have been characteristic of the different forms of selenium used or may have been due to the low selenium content of the stock diet fed to the animals (0.050 mg/kg or less).

The urinary-selenium excretion in animals suffering from chronic selenosis obviously exceeds any dietary threshold requirement, and cats poisoned with sodium selenite eliminated 50 - 80% of the intake via this pathway, but only 20% or less in the faeces (Smith et al., 1938). Cats poisoned with organic selenium in the form of seleniferous wheat protein excreted only 40% of the ingested selenium in the urine, but this decrease in urinary output was more the result of increased selenium retention rather than increased faecal excretion.

The main urinary selenium metabolite of rats is trimethyl-selenonium ion (Byard, 1969; Palmer et al., 1969). This form accounts for 20 - 50% of the urinary selenium, regardless of the form of selenium given (Palmer et al., 1970). A recent report indicates that trimethylselenonium ion accounts for a greater fraction of urinary selenium under conditions of high selenium exposure than under conditions of low exposure (Nahapetian et al., 1983). When selenate is injected into rats, over 35% of the urinary selenium is in an inorganic form, but less than 3% of the urinary selenium is inorganic when selenomethionine is injected into rats. A major urinary metabolite that accounted for 11 - 28% of the total selenium in rat urine was not identified.

The faecal excretion of selenium is more important in ruminants than in non-ruminants. For example, sheep given radio-selenite, orally, excreted 66% of the dose in the faeces, whereas swine excreted only 15% of the dose via this route (Wright & Bell, 1966). This increased faecal excretion of selenium by ruminants is the result of poor absorption rather than elevated endogenous excretion. The selenite is thought to be reduced to insoluble or unavailable forms by rumen microorganisms.

The results of studies on rats have demonstrated that
excretion of selenium by the exhalation of volatile compounds
is quantitatively significant only when animals are injected
with doses approaching almost lethal levels of soluble
selenium salts (Olson et al., 1963). A close dose-effect
relationship between selenium exhalation and selenium exposure
was observed, since the amount of selenium exhaled decreased
as the amount of selenium administered decreased. Rats fed
toxic levels of selenium in the diet on a long-term basis,
either as seleniferous wheat or sodium selenite, exhaled less
than 2% of the selenium ingested.

6.3.2 Human studies

6.3.2.1 Excretion of selenium

Human volunteers dosed orally with microgram quantities of
selenite or selenomethionine excreted 3 - 4 times more
selenium in the urine than in the faeces over a 2-week
collection period (Table 13). Subjects given selenite
excreted roughly twice as much total selenium as subjects
given selenomethionine, but the urinary pathway was dominant
in both cases. As might be anticipated from the low doses
administered, losses of selenium via the dermal or pulmonary
routes were negligible.

Table 13. Urinary and endogenous faecal excretion of radio-selenium
by human beings over a 2-week period[a]

Labelled selenium compound	Urinary excretion	Faecal excretion (endogenous)
	% of absorbed dose	
selenite	18	5
selenomethionine	9	2

[a] Adapted from: Griffiths et al. (1976).

Urinary excretion assumes greater importance when human
beings ingest solutions of milligram quantities of selenium
compounds that are soluble and also well absorbed (Table 14)
(section 6.1.1.2). Volunteers given one milligram of selenium
as selenate excreted 81% in the urine, over 3 times that
excreted after similar sized doses of selenite-selenium, and
was still twice as high when expressed as % absorbed dose.
Total excretions in urine and faeces for both selenite-

Table 14. Urinary and faecal excretion of one milligram
doses of selenium by human volunteers during a 5-day period[a]

Selenium compound	Faecal excretion % dose	Urinary excretion	
		% dose	% absorbed dose
Selenite	40	23	40
Selenate	8	81	87
Selenomethionine	6	18	20

[a] Adapted from: Thomson & Robinson (1986) (section 6.1.1.2).

selenium (63%) and selenate-selenium (90%) suggested that
long-term retention of selenium was not high. Much less of
one milligram selenium as selenomethionine was excreted in
urine and faeces (24% dose), which indicated that selenomethionine was mainly retained in comparison with selenate-
and selenite-selenium. No pulmonary excretion was observed in
one subject given one milligram of selenium as selenite
(Thomson, 1974).

Urinary-selenium levels have long been used to monitor
occupational selenium exposure (Glover, 1970), but little is
known about the urinary excretion of selenium derived from
foods. The results of balance studies conducted in New
Zealand suggested that roughly half of the dietary intake of
selenium was excreted in the urine (Robinson et al., 1973;
Stewart et al., 1978). However, these subjects were ingesting
small amounts of selenium (18 - 34 µg/day), and this
relationship may not hold for persons with higher selenium
intakes.

The importance of the kidneys in the homeostasis of
selenium was emphasized by work from New Zealand that
demonstrated that persons of low selenium status had low renal
plasma clearances of selenium and excreted selenium more
sparingly than others (Robinson et al., 1985).

6.3.2.2 Balance studies

Metabolic trials, conducted on 4 New Zealand female
volunteers in selenium balance, and consuming daily 18 - 34
µg of selenium occurring naturally in the diet, showed that
roughly equivalent amounts of selenium were voided in the
urine and faeces. In the USA, a controlled depletion/repletion
study carried out on 6 healthy young male volunteers in a
metabolic unit (Levander et al., 1980) indicated the rapidity

with which both the urinary and faecal routes adjust to differences in selenium intake (Fig. 4, 5). When the subjects were brought into the metabolic ward and given a low-selenium formula diet providing 19 - 24 µg selenium/day (depletion period), urinary-selenium excretion started to fall immediately and decreased from an initial level of 54 ± 11 µg/day to 29 ± 5 µg/day within 14 days. Faecal selenium output also responded rapidly and declined from 33 ± 13 µg/day to 17 ± 7 µg/day, after only 3 days. During the time that the subjects were on the low-selenium diet and were in negative selenium balance, urinary excretion accounted for about 63% of the total selenium output (Table 15). When the subjects were fed the low-selenium formula diet plus additional selenium in the form of seleniferous wheat and/or tuna fish (203 - 224 µg selenium/day), both urinary and faecal selenium excretion increased rapidly. During this repletion period, the subjects were in positive selenium balance and urinary excretion comprised about 45% of the total selenium output. Thus, despite a 6-fold difference in selenium intake and a transition from negative to positive selenium balance, the proportion of selenium excreted in the urine remained relatively constant in terms of the total selenium output, when the selenium was supplied as it occurs naturally in foods.

The amount of selenium excreted in the sweat was found to be very low, and was not affected by changes in dietary-selenium intake (Table 16). In agreement with these results, Griffiths et al. (1976) found very little radio-selenium in the sweat of subjects who had been given labelled selenite or selenomethionine. In the study by Levander et al. (1980), salivary-selenium levels were also very low, but appeared to reflect differences in dietary-selenium intakes, since salivary-selenium was somewhat higher at the end of the repletion period than at the end of the depletion period (Table 16). Hadjimarkos & Shearer (1971) found 1.1 - 5.2 µg selenium/litre of saliva in normal children.

6.4 Retention and Turnover

6.4.1 Animal studies

Results of studies on rats have shown that the whole-body retention of a single injected dose of radioactive selenite is described by a curve that consists of an initial phase, a transition phase, and an extended phase (Ewan et al., 1967; Burk et al., 1972). During the initial equilibration phase, there is a distribution of radioactive selenium throughout the tissues and a rapid excretion of radio-selenium in the urine, faeces, and, if the dose is sufficiently high, the expired

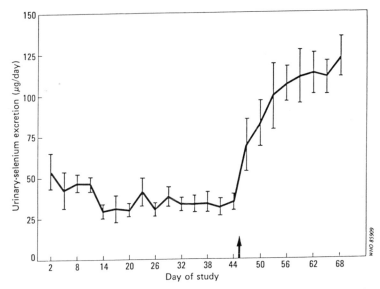

Fig. 4. Effect of experimental selenium depletion and repletion on the urinary excretion of selenium by young male volunteers. Arrow indicates end of depletion and start of repletion. Values are means ± standard deviation. From: Levander et al. (1980).

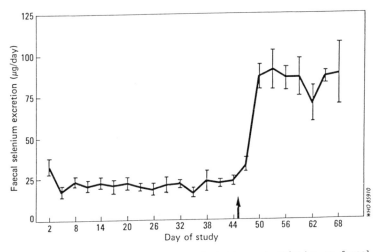

Fig. 5. Effect of experimental selenium depletion and repletion on faecal excretion of selenium by young male volunteers. Arrow indicates end of depletion and start of repletion. Values are means ± SD. From: Levander et al. (1980).

Table 15. Selenium balance during depletion and repletion periods[a]

	Selenium balance	
	Depletion period	Repletion period
	(µg)	(µg)
Urinary excretion	-1665 ± 211	-2435 ± 272
Faecal excretion	-985 ± 103	-1919 ± 228
Total excretion	-2650 ± 257	-4354 ± 295
Intake	+1524 ± 173	+5424 ± 248
Overall balance	-1126 ± 268	+1070 ± 482

[a] Data expressed as mean ± standard deviation.
Adapted from: Levander et al. (1981b).

Table 16. Effect of selenium depletion and repletion on the selenium contents of saliva and sweat in young men[a]

Body fluid	Selenium content		
	Start of study	End of depletion	End of repletion
	(µg selenium/litre)		
Saliva	2.8 ± 0.7[b,c]	1.4 ± 0.4[b]	4.4 ± 0.4[c]
Sweat	-	1.4 ± 0.2	1.2 ± 0.7

[a] Adapted from: Levander et al. (1981b).
[b,c] Means in the same row with different superscript letters differ significantly ($P < 0.05$, Duncan's multiple range test). Data expressed as mean ± SD.

air. The initial phase is followed by a transition phase, during which the rate of loss of radio-selenium is less than that in the initial phase but greater than that in the final extended phase. The extended phase consists of a slow constant rate of radio-selenium loss which represents the long-term whole-body turnover of selenium. Increasing the dose of radioactive selenite administered decreased the percentage of the dose retained during the initial phase, but did not have any effect on radio-selenium retention during the extended phase. Increasing the level of dietary selenium increased the turnover of radio-selenium during both the initial and extended phases. For example, supplementing the diet with 0.95 mg selenium/kg decreased the biological half-life of radio-selenium, during the extended phase, from 78 to 27 days. The extended phase of radio-selenium retention in rats was shown by Richold et al. (1977) to be largely independent of the route of administration and the chemical form of selenium given. Thus, after the initial period, it appears that selenium from a variety of sources is ultimately incorporated into the same metabolic pool in rats. The apparent "whole-body" retention of radio-selenium during the extended phase is an average of several discrete processes, since each internal organ has its own characteristic rate of selenium turnover. For example, the biological half-times of radio-selenium in the kidneys, whole body, and skeletal muscle of rats were 38, 55, and 74 days, respectively (Thomson & Stewart, 1973).

Information on the retention of selenium by animals became of practical significance when approval was sought for the use of selenium as a feed additive to prevent nutritional deficiency diseases (section 4). Because of the toxicity of high levels of selenium, there was some concern about the possible build up of residues in the edible flesh of animals supplemented with selenium. Scott & Thompson (1971) noted very little increase in the concentrations of selenium in the tissues of chicks or poults fed low-selenium diets supplemented with 0.2 - 0.8 mg selenium/kg as sodium selenite. However, feeding a diet naturally high in selenium, but not toxic (0.67 mg/kg) resulted in substantially increased tissue-selenium levels.

Scott & Thompson (1971) also showed that addition of sodium selenite to a diet that was naturally high in selenium did not produce any further increase in tissue-selenium levels. In further studies on turkeys, swine, sheep, and cattle, animal feeds, naturally low or marginal in selenium, were supplemented with nutritional levels of selenium (0.1 - 0.5 mg/kg) as sodium selenite. Resulting tissue-selenium concentrations were not any higher than those that would be expected if the animals were fed diets naturally adequate in selenium (Groce et al., 1971; Cantor & Scott, 1975; Ullrey et al., 1977).

These observations are in agreement with those from earlier work with toxic levels of selenium, which indicated that naturally-occurring organically-bound selenium was retained in the tissues to a much greater extent than inorganic selenium (Smith et al., 1938). Martin & Hurlbut (1976) fed mice high levels of selenium as selenite, selenium-methylselenocysteine, or selenomethionine. After 7 weeks, the mice fed selenomethionine had much higher levels of selenium in their tissues than those fed either selenite or selenium-methylselenocysteine. Moreover, when the various selenium compounds were removed from the diet, selenium was retained more strongly in the tissues of the mice fed the seleno-methionine than in those of the mice fed selenite or selenium-methylselenocysteine. The results of this study showed that a distinction more precise than inorganic versus organic must be made when discussing the metabolism of selenium compounds, since the metabolism of selenium-methylselenocysteine resembled that of selenite rather than that of seleno-methionine.

Differences in the retention of selenite compared with selenomethionine were also observed when these 2 forms of selenium were fed in the diet at nutritional levels. For example, Miller et al. (1972) showed that the total body selenium content of chicks was increased by an average of 29.1%, when 0.025 - 0.500 mg of selenium as selenomethionine was fed per kg of diet, but was increased only by 17.9% when selenium as selenite was fed at similar levels.

Selenium retained by female animals can apparently be used later to protect their offspring against the effects of selenium deficiency. Allaway et al. (1966) fed ewes alfalfa containing 2.6 mg selenium/kg for 5 months followed by a diet low in selenium. The ewes were able to transmit levels of selenium to their lambs that protected them against White Muscle Disease, even though the lambs were born 10 months after the low-selenium diet was first administered.

6.4.2   Controlled human studies

As in the case of rats, the total body retention curves of radioactive selenium in human beings can be resolved into 3 components (Griffiths et al., 1976). However, oral doses of ^{75}Se-selenomethionine are retained more strongly and turned over more slowly than oral doses of ^{75}Se-selenite (Table 17). This metabolic difference between selenomethionine and selenite holds true, even if the compounds are administered by intravenous injection (Lathrop et al., 1972; Falk & Lindhe, 1974). These results differ with those obtained in rats in which several forms of selenium apparently were incorporated into the same long-term metabolic pool and were thus turned over at similar rates (Richold et al., 1977). On this basis, it was suggested that selenomethionine, or food-selenium in a form that produced selenomethionine after digestion, might prove more effective than selenite in improving a low-selenium status or in correcting selenium deficiency in man (Griffiths et al., 1976).

Thomson et al. (1982) gave 100 µg doses of selenium as selenomethionine or sodium selenite to 12 New Zealand volunteers over a period of several weeks. Blood glutathione

Table 17. Retention of oral doses of selenite and selenomethionine
by human beings[a]

Form of selenium given	Whole-body selenium retention 14 days after dose (% of absorbed dose)	Biological half-times for whole-body selenium retention		
		phase 1	phase 2 (days)	phase 3
selenite	78	1.0	8	103
seleno-methionine	89	1.7	18	234

[a] Adapted from: Griffiths et al. (1976).

peroxidase (EC 1.11.19) activity increased in all subjects
and, at 17 weeks, the response was similar in both selenomethionine and selenite groups. Increases in blood-selenium
levels were greater after supplementation with selenomethionine than with selenite. Selenium levels tended to
plateau in the selenite-treated subjects but kept increasing
in the selenomethionine-treated subjects. Similar differences
in the retention and turnover of selenium as selenate compared
with selenium in selenium-rich yeast or wheat were observed in
a bioavailability trial in Finland (Levander et al., 1983)
(section 5.1.1.6). Thus, it appears that feeding of selenium
bound in organic form (selenomethionine, selenium-rich yeast
or wheat) results in higher blood-selenium levels than the
feeding of selenium as selenate or selenite, but there is no
difference in the glutathione peroxidase activities ultimately
achieved during supplementation. However, when the selenium
supplements were discontinued, the glutathione peroxidase
activities remained somewhat elevated in the groups receiving
the wheat or yeast compared with those receiving selenate.
Apparently, the selenium in the yeast or wheat retained in the
tissues could be used for glutathione peroxidase production
after the selenium supplement was discontinued.

6.5 Metabolic Transformation

Some metabolic transformations of selenium compounds are
outlined in Fig. 6. This scheme tends to emphasize the
central role of selenite, but other forms of selenium may have
important characteristic pathways of their own, under certain
conditions (e.g., the direct incorporation of selenomethionine
as such into tissue proteins).

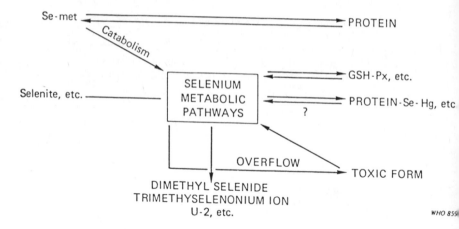

Fig. 6. Schematic diagram of selenium metabolism showing possible pathways of incorporation of dietary selenium (on left) to tissue selenium (on right). From: Burk (1976).

6.5.1 Animal studies

6.5.1.1 Reduction and methylation

The main flow of selenium metabolism in animals is via reductive pathways, which contrasts with the primarily oxidative metabolism of sulfur (Levander, 1976a). Selenite can react with glutathione or protein sulfhydryls to form selenotrisulfides (Ganther, 1968; Jenkins & Hidiroglou, 1971) which, at least in the case of the glutathione derivatives, can be reduced further by an enzymatic mechanism to selenide (Ganther & Hsieh, 1974; Diplock, 1976).

Under normal conditions, selenide is methylated to form trimethylselenonium ion, the main urinary metabolite of selenium (section 6.3.1). In cases of selenium toxicity, this pathway is overloaded and dimethyl selenide is produced. This is the main volatile selenium metabolite expired via the lungs and is responsible for the typical "garlicky odour" of animals poisoned with selenium (McConnell & Portman, 1952a). The last 2 reactions could be considered detoxication steps, since the methylated end-products are much less toxic for the organism than selenite (McConnell & Portman, 1952b; Obermeyer et al., 1971). However, both of these methylated selenium derivatives have strong synergistic toxicity with other minerals (section 7) and dimethyl selenide toxicity in male rats was reported to be inversely related to the level of previous selenium intake (Parizek et al., 1980).

6.5.1.2 Form in proteins

Early work with animals poisoned with selenium revealed
that much of the selenium in the tissues was associated with
protein (Smith et al., 1938). Since that time there has been
considerable controversy about the exact chemical nature of
the selenium in tissue proteins (Levander, 1976a). Selenomethionine may be incorporated initially as such into animal
proteins (Ochoa-Solano & Gitler, 1968), but, in rats, it
appears to be eventually catabolized to selenite or selenate
(Millar et al., 1973; Thomson & Stewart, 1973). Non-ruminant
animals cannot synthesize selenomethionine from inorganic
selenium compounds (Cummins & Martin, 1967; Olson & Palmer,
1976), but rabbits and rats can convert selenite into selenocysteine tissue proteins (Godwin & Fuss, 1972; Olson & Palmer,
1976). Forstrom et al. (1978) have proposed that selenocysteine is the form of selenium located at the catalytic site
of glutathione peroxidase and others have shown that selenocysteine is essential for the activity of clostridial glycine
reductase (Cone et al., 1976). The mechanism by which selenocysteine is formed from selenite is not known but incorporation of preformed selenocysteine and post-translational
modification of the protein have been considered (Sunde,
1984). Other possible forms of selenium in proteins include
selenotrisulfides (Ganther & Corcoran, 1969) and "acid-labile"
selenium (Diplock et al., 1973).

6.5.1.3 Conversion of selenium compounds to nutritionally-
active forms of selenium

The pioneering studies of Schwarz & Foltz (1958) demonstrated that a variety of selenium compounds could protect
against dietary liver necrosis in vitamin E- and selenium-
deficient rats. Sodium selenite, sodium selenate, selenium
dioxide, selenic acid, and potassium selenocyanate were all
more or less equally active, but elemental gray selenium was
essentially inactive. Selenocystine and selenomethionine were
about as effective as the active inorganic selenium compounds,
but an organic selenium fraction isolated from pig kidney
powder (Factor 3) was shown to be even more active. A number
of organic derivatives of selenium have shown some protective
effect against liver necrosis, but none was superior to Factor
3 (Schwarz et al., 1972). However, certain simple amino-acid
derivatives of monoseleno diacetic acid showed some promise in
that they combined high nutritional potency with a low order
of toxicity (Schwarz, 1976). The effects of selenite and
selenomethionine were shown to be roughly similar in inducing
glutathione peroxidase activity in the tissues of rats fed a
diet deficient in selenium (Pierce & Tappel, 1977).

Cantor et al. (1975a) found that while selenite and selenocystine were equally effective in preventing exudative diathesis in vitamin E- and selenium-deficient chicks, seleno-methionine was less effective. This suggests a difference between rats and chicks in the metabolism of selenomethionine. The degree of protection against exudative diathesis and the level of plasma glutathione peroxidase activity were highly correlated, suggesting that nutritional potency depended on the ability of the chick to convert various selenium compounds to the enzymatically active form.

6.5.2   Human studies

Few studies have been carried out to investigate the metabolic transformation of selenium compounds in human beings, but the limited data available suggest certain similarities in the metabolism of selenium by man and animals. For example, human beings overexposed to selenium develop breath with a smell of garlic, which is presumably due to the exhalation of dimethyl selenide (Glover, 1976). Also, the chromatographic pattern of urinary-selenium metabolites is the same in man and the rat (Burk, 1976). However, species differences in selenium metabolism do exist, since human beings retain seleno-methionine to a much greater extent than selenite, whereas the retention of the 2 compounds in rats is about the same (section 6.4.2).

7. EFFECTS OF SELENIUM ON ANIMALS

The considerable biological importance of selenium was first recognized in the 1930s when it was discovered that certain well-defined and economically important farm animal diseases were actually the result of chronic selenium poisoning (section 7.1). These animal diseases were restricted to agricultural areas in which large amounts of selenium in the soil were available for uptake by the plants, which were then consumed by the animals. Research, over the last 20 years, showed that selenium was an essential trace element (section 7.2), and selenium deficiency diseases were rapidly recognized in several species of farm animals. These deficiency diseases are significant economic problems in areas of the world where the soil levels of the element available for uptake by plants are low.

7.1 Selenium Toxicity

7.1.1   Farm animal diseases associated with a high selenium intake

On the basis of field experience, Rosenfeld & Beath (1964) delineated 3 different types of selenium poisoning in livestock:

(a)   acute;

(b)   chronic, of the blind staggers type; and

(c)   chronic, of the alkali disease type.

Acute poisoning is due to the ingestion of toxic quantities of selenium in the form of highly seleniferous accumulator plants. The animal has severe signs of distress such as laboured breathing, abnormal movement and posture, prostration, and diarrhoea. Death often follows within a few hours. This type of selenium poisoning is rather rare under field conditions, since grazing animals generally avoid the selenium accumulator plants, except in times of pasture shortage. Acute selenium poisoning has also been produced by the experimental or accidental administration of selenium compounds to farm animals (US NAS/NRC, 1976).

Blind staggers has been reported in animals that eat a limited number of selenium accumulator plants over a period of weeks or months (Rosenfeld & Beath, 1964). The affected animals wander, stumble, have impaired vision, and eventually succumb to respiratory failure. Although this type of poisoning can be produced experimentally by the administration of water extracts of accumulator plants, it has not been possible to duplicate this syndrome by the administration of pure selenium compounds. Possibly alkaloids or other toxic substances found in many seleniferous plants may contribute to blind staggers (Maag & Glenn, 1967; Van Kampen & James, 1978).

Alkali disease is associated with the consumption of grains containing 5 - 40 mg selenium/kg over weeks or months. Animals exhibit liver cirrhosis, lameness, hoof malformations, loss of hair, and emaciation. Maag & Glenn (1967) were unable to produce alkali disease in cattle by feeding inorganic selenium, but several other studies have shown that the syndrome is causally associated with seleniferous grains or grasses and can be produced by feeding inorganic selenium salts (Olson, 1978).

The Task Group noted that most of the work on alkali disease was concerned with cattle, but was aware of the report by Ermakov & Kovalskij (1968), which described chronic selenium toxicity of this type in sheep, under natural conditions. In sheep fed feeds containing levels of 2 mg selenium/kg feed (fresh weight), the following characteristic signs were observed: hoof deformation, loss of hair, hypochromic anaemia, and increases in the activity of both alkaline and acid phosphatases in various tissues.

7.1.2   Toxicity in experimental animals

The practical significance of selenium poisoning in farm animals stimulated a great deal of research on both the acute and chronic effects of selenium in laboratory animals. Interest in the toxic effects of repeated exposure to selenium via inhalation was stimulated by concern about the possible effects on human health of occupational exposure to selenium. Also, studies were carried out with the aim of establishing the no-observed-adverse-effect dose level of selenium when administered in the drinking-water. In some of the studies in which selenium was given via either the air or water, biochemical and/or behavioural criteria were used to assess the biological effects of selenium exposure. However, as discussed in section 7.1.6, the toxicity of selenium compounds can be influenced by different variables and the results from various laboratories are often not comparable because of quite different experimental conditions. Also, the criteria for toxicity are less developed than the signs of deficiency (section 7.2).

Reviews that deal with various aspects of selenium toxicology include those by Rosenfeld & Beath (1964), Muth (1966), Izraelson et al. (1973), Ermakov & Kovalskij (1974), US NAS/NRC (1976), and Lazarev (1977).

7.1.2.1 Acute and subacute toxicity - single or repeated exposure studies with oral, intraperitoneal, or cutaneous administration

Perhaps the most characteristic sign of acute selenium poisoning in animals is the development of the so-called "garlicky breath odour", which is due to the pulmonary excretion of volatile selenium compounds, particularly dimethyl selenide, by animals overexposed to selenium (section 6.3.1). Other signs of acute selenium poisoning described by Franke & Moxon (1936) in dogs and rats included: vomiting, dyspnoea, tetanic spasms, and death from respiratory failure. Pathological changes included congestion of the liver with areas of focal necrosis, congestion of the kidney, endocarditis, myocarditis, peticheal haemorrhages of the epicardium, atony of the smooth muscles of the gastro-intestinal tract, gall-bladder, and bladder, and erosion of the long bones, especially the tibia. The LD_{50} values for sodium selenite, administered orally to various animal species, are given in Table 18 (Pletnikova, 1970).

Table 18. Acute oral toxicity of sodium selenite for various species of laboratory animals[a]

Species	LD_{50} (mg selenium/kg body weight)	Statistical method
White mouse (male)	7.75	Behrens & Schlosser
	7.08	Litchfield & Wilcoxon
Albino rat (female)	10.50	Behrens & Schlosser
	13.19	Diechmann & LeBlanc
Guinea-pig (female)	5.06	Deichmann & LeBlanc
Rabbit (female)	2.25	Diechmann & LeBlanc

[a] From: Pletnikova (1970).

For a given species, the lethal doses of sodium selenite, sodium selenate, DL-selenocystine, and DL-selenomethionine are quite similar (Table 19). Although certain methylated metabolites of selenium such as dimethylselenide and trimethylselenonium chloride were considered relatively innocuous (see LD_{50} values by McConnell & Portman and Obermeyer et al. in Table 19), more recent work by Parizek et al. (1976, 1980) has shown that the toxicity of methylated selenium compounds depends not only on the sex of the animal

Table 19. Acute toxicity of some selenium compounds administered to
rats by intraperitoneal injection

Compound	Criterion of toxicity	Toxic dose (mg selenium/kg body weight)	Reference
Sodium selenite	MLD	3.25 - 3.5	Franke & Moxon (1936)
Sodium selenate	MLD	5.25 - 5.75	Franke & Moxon (1936)
DL-selenocystine	MLD	4	Moxon (1940)
DL-selenomethionine	MLD	4.25	Klug et al. (1950)
Diselenodipropionic acid	LD_{50}	25 - 30	Moxon et al. (1938)
Trimethylselenonium chloride	LD_{50}	49.4	Obermeyer et al. (1971)
Dimethyl selenide	LD_{50}	1600	McConnell & Portman (1952b)

(Parizek et al., 1974) but also on the level of previous
selenium intake. For example, 90% mortality was observed in
male rats maintained on a diet containing 0.05 mg selenium/kg,
when they were injected intraperitoneally with 20 μmoles
dimethylselenide/kg body weight, i.e., by a dose of dimethylselenide that is more than 1000 times lower than the LD_{50}
reported by McConnell & Portman (1952b). Pre-treatment with a
small amount of selenite (1 μmol/kg body weight), intraperitoneally, 6 h, but not 1 h, before dimethylselenide
injection or increased oral intake of selenite (Table 20),
protected male rats against the toxicity of a subsequent dose
of dimethylselenide (Parizek et al., 1976, 1980). Moreover,
the methylated forms of selenium have strong synergistic
toxicities with other minerals (section 7.1.6.3, 7.4.1) and
dimethylselenide can be much more toxic for male rats than for
female rats (section 7.1.6).

Smith et al. (1937) reported that the minimum lethal dose
of selenium as sodium selenite or selenate in rabbits, rats,
and cats was 1.5 - 3.0 mg/kg body weight, regardless of
whether the compounds were administered orally, subcutaneously, intraperitoneally, or intravenously. This lack of
effect of the mode of administration probably reflects the
rapid and complete absorption of soluble selenium compounds,
either from the site of injection or from the gastrointestinal
tract.

Cummins & Kimura (1971) described comparative studies on
Sprague Dawley rats and dogs concerning the oral toxicity of
the following selenium compounds: sodium selenate, selenourea,
biphenylselenium, selenium sulfide (1 - 30 μ particle size),
and elemental selenium (1 - 30 μ particle size). The oral
LD_{50} values in rats for a number of selenium compounds are

Table 20. Dependence of the toxicity of dimethylselenide on the level of previous oral intake of selenium[a]

Selenite supplement in drinking-water (mg selenium/ litre	Single intraperitoneal dose of dimethylselenide (μg mole/kg body weight)	24-h mortality (%) Diet A (n = 20)	Diet B (n = 10)
0	20	90	90
0.1	20	45	30
0.5	20	5	0
1.0	20	0	0

[a] Adapted from: Parizek et al. (1976, 1980).

Male rats (2 months old) given drinking-water with stated supplement of selenite for 3 days before dimethylselenide administration. Diet A contained 0.052 ± 0.005 mg selenium/kg and diet B (semi-synthetic diet) 0.044 ± 0.001 mg selenium/kg.

shown in Table 21 and demonstrate the large variations in LD$_{50}$ values that occur, depending on both the oxidation state of the compound and its aqueous solubility.

The aqueous solubilities were carried out in 0.01 N HCl to more closely simulate acidic conditions in the stomach. The least toxic selenium compound was insoluble elemental selenium with an LD$_{50}$ of 6.7 g/kg body weight. Toxic signs included pilomotor activity, decreased body activity, dyspnoea, diarrhoea, anorexia, and cachexia. Fatalities occurred within 18 - 72 h; survivors appeared outwardly normal, at the end of the 7-day observation period. The most toxic of the selenium compounds tested was the highly soluble sodium selenite with an oral LD$_{50}$ of 7 mg/kg body weight and the toxic signs were similar to those seen with high doses of elemental selenium. Selenium sulfide (a component in shampoos) was about 20 times less toxic than sodium selenite (i.e., an LD$_{50}$ of 138 mg/kg compared with 7 mg/kg). It was also found in this study that, with the exception of biphenyl selenium, toxicity could be correlated with blood-selenium levels. For example, sodium selenite being the most toxic, gave the highest blood level followed in descending order by selenourea, selenium sulfide, and elemental selenium. It was suggested that the blood-selenium level produced by the relatively non-toxic biphenyl selenium was high because this covalently-bound selenium compound is the most lipophilic compound, which is better absorbed, resists catabolism, and apparently circulates as the parent compound.

Table 21. Comparative solubility and toxicity of various
selenium compounds in rats[a]

Compound	Solubility in 0.01 N HCl	Rat oral LD_{50} (95% CL) (mg/kg body weight)
Na_2SeO_3	700 g/litre	7 (4.4 - 11.2)
$H_2N-C-NH_2$ (with $\overset{\|}{Se}$)	30 g/litre	50 (35.7 - 70.0)
SeS_2	insoluble (< 1 g/litre)	138 (110 - 172)
Se	5 g/litre	360 (308 - 421)
Elemental Se	insoluble (< 1 g/litre)	6700 (6000 - 7300)

[a] From: Cummins & Kimura (1971).

Male Sprague Dawley rats, each weighing between 50 and 100 g, were given the above chemicals by gavage as 0.1 - 20% suspensions in 0.5% methylcellulose. A total of 30 - 36 animals was used per compound, in groups of 6 animals per dose. The LD_{50} values and the associated confidence limits (CL) were calculated according to the method of Litchfield & Wilcoxon.

The application of 83 mg of selenium oxychloride to the skin of rabbits caused the death of the animals in 5 h, and application of 4 mg caused death in 24 h (Dudley, 1938). Lazarev (1977) reported a minimum lethal dose of selenium oxychloride of 7 mg/kg body weight, after cutaneous administration to rabbits.

Several papers have shown that injection of selenite in rats in the early post-natal period in single doses of 1.5 mg selenium/kg body weight, or in repeated doses of 0.5 mg/kg, induced cataracts (Ostadalova et al., 1979; Bhuyan et al., 1981; Shearer et al., 1980; Bunce et al., 1985). Similar effects have not been observed in hamsters (Shearer et al., 1980). This response is dose dependant (Table 22) and, thus far, has been observed only when the selenite was given by parenteral injection.

7.1.2.2 Effects of long-term oral exposure

Moxon & Rhian (1943) summarized several older studies that indicated that diets containing 5 mg selenium/kg or more cause

Table 22. Cataractagenesis by selenite in the rat[a]

Group	Daily dosage[b] (mg selenium/kg body weight)	Frequency of cataracts[c] (%)
Control	0	0
Na_2SeO_3	0.25	13
Na_2SeO_3	0.50	96
Na_2SeO_3	0.75	96

[a] Adapted from: Shearer et al. (1980).
[b] Dose given to rat pups daily on days 2 - 18 post partum.
[c] Observed at 21 days of age.

chronic selenosis in several species of animals, such as chickens, rats, and dogs. In seleniferous areas, 5 mg selenium/kg diet is generally accepted as the dividing line between toxic and non-toxic feeds (US NAS/NRC, 1976).

Halverson et al. (1966) fed diets containing 0, 1.6, 3.2, 4.8, 6.4, 8.0, 9.6, or 11.2 mg selenium/kg in the form of sodium selenite or seleniferous wheat to male Sprague Dawley rats, initially weighing 60 - 70 g. The rats were divided into groups of 8 and were fed the diets for 6 weeks. The diet consisted of (g/kg) ground wheat, 809; purified casein 120; USP salt mixture XIV, 20; USP brewer's yeast, 20; corn oil, 30; and vitamin B_{12} mix, 1. Vitamins A, D, and E were provided separately. The criteria used to assess toxicity were growth depression, liver cirrhosis, splenomegaly, pancreatic enlargement, anaemia, elevated serum-bilirubin levels, and death. The addition to the diet of 1.6, 3.2, or 4.8 mg selenium/kg did not have any significant effect on the rats, as judged by any of these criteria. There was a depression in growth in the group that received 4.8 mg dietary selenium/kg as sodium selenite, but this was not significant. Liver cirrhosis, splenomegaly, and significant growth depression were observed when the rats were fed levels of 6.4 mg/kg or more from either source of selenium. Diets containing 8.0 mg/kg or more caused additional effects such as pancreatic enlargement, anaemia, elevated serum-bilirubin levels, and, after 4 weeks, death.

Diets made either with non-seleniferous or seleniferous sesame meal were fed to groups of 12 male Sprague Dawley rats, initially weighing 50 g for a period of 6 weeks (Jaffe et al., 1972b). The diet made with non-seleniferous sesame meal contained 0.5 mg selenium/kg and consisted of (g/kg): non-seleniferous sesame meal, 463.3; almidon, 422.7; corn oil, 50; cod liver oil, 10; USP XVI salts, 40; L-lysine•HCl, 4; and vitamin mix, 10. The diet made with the seleniferous sesame meal contained 10 mg selenium/kg and had the same composition as the previous diet except that seleniferous sesame, which replaced the non-seleniferous sesame meal, was added at a level of 490.2 g/kg, and the almidon was added at a level of 395.8 g/kg. The rats fed the diet containing the seleniferous sesame meal showed decreased survival, impaired weight gain, higher incidence of liver lesions, elevated hepatic selenium levels, enlarged spleens, depressed haemoglobin, haematocrit, and fibrinogen levels, and decreased prothrombin activity (Table 23). In a separate study, the same workers investigated the effects of dietary selenium, given as seleniferous sesame meal, on the activities of various serum enzymes (Table 24). The diet containing 4.5 mg selenium/kg had the same composition as the previous diets, except that the seleniferous sesame meal, non-seleniferous sesame meal, and almidon were added at levels of 230.8, 270.4, and 384.8 g/kg, respectively. Feeding the diet containing 4.5 mg of selenium/kg for 6 weeks increased the activities of serum alkaline phosphatase (EC 3.1.3.1) and glutamic-pyruvic transaminase (SGPT) (EC 2.6.1.2), whereas 10.0 mg/kg also increased the activity of glutamic-oxaloacetic transaminase (SGOT) (EC 2.6.1.1).

Tinsley et al. (1967) and Harr et al. (1967) carried out an extensive study on the chronic toxicity of selenium in rats. Although the primary purpose of their research was to investigate the alleged carcinogenicity of selenium (section 7.7.1), the design of their study provided an opportunity to examine other aspects of the toxicity of selenium, such as the influence of selenium poisoning on growth rate and histopathology. A total of 1437 Wistar rats from a closed random-bred colony was used with the size of the experimental groups ranging from 10 - 110 animals. The rats were fed one of 3 different diets: a semipurified diet containing either 12 or 22% casein or a commercial "laboratory chow" type ration. The casein-based diets also contained corn oil, 50 g; H.M.W. salts, 40 g; vitamin mix, 10 g; and glucose monohydrate (Cerelose) up to a weight of 1 kg. Selenium as sodium selenite or sodium selenate was added at levels of 0, 0.5, 2.0, 4.0, 8.0, or 16.0 mg/kg. Only a small proportion of the animals fed diets supplemented with more than 4.0 mg/kg of selenium survived for 12 months. The rats fed the commercial ration were 2 - 3 times more resistant to the effects of selenium toxicity than the rats fed the semipurified diet. A calculated maximum body weight was reported to be depressed by as little as 0.5 mg selenium/kg, but no statistical evaluation of the results was presented. Usually moribund animals were

Table 23. Pathological signs in rats fed diets containing seleniferous sesame meal[a]

Dietary selenium	Survival after 6 weeks	Weight gain after 6 weeks	Rats with hepatic lesions	Hepatic selenium level	Spleen weight	Haemoglobin level	Haematocrit value	Fibrinogen	Prothrombin activity
(mg/kg)		(g)		(mg/kg)	(% of body weight)	(g/litre blood)	(volume %)	(mg/litre)	
0.5	12/12	156.8±7.2	0/12	0.72±0.14	0.207±0.01	147±2.1	43.9±0.61	1664±78	963±9.5
10	8/12	61.2±7.43	10/12	7.34±0.97	0.544±0.09	123±3.5	40.8±0.82	655±78	713±76.6

[a] Adapted from: Jaffe et al. (1972b).

Table 24. Effect of chronic selenium toxicity on the
activity of serum enzymes[a]

Dietary selenium	Number of rats	Alkaline phosphatase	SGOT	SGPT
(mg/kg)		(units/ml)	(units/ml)	(units/ml)
0.5	6	6.3 ± 0.2	66 ± 3	20 ± 1
4.5	10	8.8 ± 1.3	59 ± 2	26 ± 2
10	8	18.9 ± 2.3	77 ± 5	43 ± 6

[a] Adapted from: Jaffe et al. (1972b).

killed and necropsied for histopathological determinations,
but 136 rats were killed at specific ages.

Acute toxic hepatitis was the predominant histopathological lesion in rats fed the commercial ration supplemented
with sodium selenate at 16 mg selenium/kg. These rats had a
median survival age of only 96 days. Acute toxic hepatitis
was also observed in rats fed the 12% casein diet supplemented
with sodium selenate at 4 or more mg selenium/kg. The rats
were emaciated, pale, and had poor quality hair coat.
Hydrothorax, ascites, pericardial oedema, and icterus were
common. Myocardial hyperaemia, fluid imbalance, and
parenchymal degeneration were often present. The adrenals
were enlarged and the pancreas was oedematous. The failure of
normal chondrocyte proliferation observed in the metaphyses
appeared to be different from that seen in malnutrition, since
proliferation was irregular and not merely reduced.

When rats were fed the commercial ration supplemented with
sodium selenate at 8 mg selenium/kg, the predominant lesion
was chronic toxic hepatitis. The median survival age in this
group was 429 days. Other histopathological changes reported
included pancreatic duct hyperplasia, intestinal nephritis,
and myocardial damage, particularly in rats of more than 450
days of age and receiving selenite.

Harr et al. (1967) reported an increased proliferation of
the hepatic parenchyma when the rats were fed the
semi-purified diet supplemented with 0.5 - 2.0 mg selenium/kg
as selenite or selenate. But a more detailed report from this
laboratory (Weswig et al., 1966) showed that this lesion of
"chronic liver and bile duct hyperplasia" was observed to a
greater extent in rats fed the commercial ration not
supplemented with selenium than in rats fed the semi-purified
diet supplemented with 0.5 mg selenium/kg. Thus, this lesion
may not be specifically related to selenium.

Harr & Muth (1972) stated that 0.25 mg/kg was the minimum toxic level for liver lesions, when selenium was added to a semi-purified diet. The minimum toxic level was 0.75 mg/kg when the criteria were longevity or lesions of the heart, kidneys, or spleen. However, growth was normal in rats fed 0.5 mg selenium/kg diet. The authors estimated that the dietary threshold for physiological and pathological effects was 0.4 mg/kg and for pathological and clinical effects, 3 mg/kg.

Weanling rats of the Long-Evans strain were given either sodium selenite or selenate at 0 or 2 mg selenium/litre in the drinking-water for 1 year (Schroeder & Mitchener, 1971b). After 1 year, the selenium dosage was increased to 3 mg/litre in the selenate group. The weanlings were born from random-bred females that had been fed a low-selenium diet (0.05 mg selenium/kg) consisting of: whole rye flour, 600 g; dry skim milk, 300 g; corn oil, 90 g; and iodized sodium chloride to which were added vitamins and iron, 1 g/kg. The same diet was fed to the weanlings during the toxicity study. The drinking-water was doubly deionized forest spring water to which had been added: zinc, 50 mg; manganese, 10 mg; chromium(III), 5 mg; copper, 5 mg; cobalt, 1 mg; and molybdenum, 1 mg/litre. The rats given the selenium compounds (plus another group not discussed here given sodium tellurite) were divided into groups totalling 313 animals. There were 105 control rats. By 58 days, half of the male rats in the selenite group were dead. Fifty percent mortality in the group of female rats given selenite was not achieved until 348 days and was not achieved in the male and female groups given selenate until 962 and 1014 days, respectively. On the other hand, Palmer & Olson (1974) gave 2 or 3 mg selenium/litre drinking-water, in the form of either sodium selenite or selenate to male weanling Sprague Dawley rats for 6 weeks and noted small decreases in weight gain compared with control rats receiving selenium, but no deaths. Two diets were used in this study, a rye diet similar to that described by Schroeder & Mitchener above and a corn diet that consisted of ground corn, 808 g; casein, 120 g; corn oil, 30 g; U.S.P. XIV salts, 20 g; and vitamin mix, 22 g/kg. The trace elements zinc, copper, manganese, chromium, cobalt, and molybdenum were also added to the drinking-water, as suggested by Schroeder & Mitchener (1971a).

In a study by Jacobs & Forst (1981a), sodium selenite at 0 or 4 mg selenium/litre drinking-water was administered to groups of 17 and 30 male 5-week-old Sprague Dawley rats fed a commercial pellet ration. After 64 weeks, survival was 94 and 63% in the control and selenium-treated groups, respectively. In a second study of similar design, except that the rats were 8 weeks old at the start, survival was 90 and 95% in the control and selenium-treated groups, respectively, after 61 weeks.

Pletnikova (1970) investigated the effects of long-term, low-level administration of sodium selenite in water to rabbits and rats. For these studies, 32 rabbits and 16 rats were divided into 4 groups and administered, orally, doses of 0, 0.005, 0.0005, or 0.00005 mg selenium/kg body weight for periods of 7 1/2 and 6 months, respectively. Prolonged administration of the maximum dose investigated (0.005 mg/kg) produced significant alterations in the rabbits. After 2 months, there was a significant increase in the concentration of oxidized glutathione in the blood and, after 7 months, there was slower elimination of bromsulphalein by the liver, and hepatic succinic dehydrogenase activity was decreased. A dose of 0.0005 mg/kg caused fewer, less pronounced changes, whereas 0.00005 mg/kg did not produce any statistically significant effects in any of these tests. A dose of 0.005 mg/kg given to rats caused a considerable weakening of the capacity for forming new conditioned reflexes. It can be assumed that a daily dose of 0.005 mg/kg body weight in rats is equivalent to a dietary-selenium level of about 0.063 mg/kg. Since this level of selenium is within the physiological range needed to prevent selenium deficiency in animals (section 7.2.2), the toxicological significance of these observations is not clear, unless it is assumed that a certain dose of selenite given in water solution is more toxic than the same dose given in the diet.

Recently, Csallany et al. (1984) reported that giving sodium selenite to female mice in the drinking-water at a level of 0.1 mg selenium/litre increased the amount of hepatic lipid-soluble lipofuscin pigments, when the animals were sacrificed at 9 months of age. There were 8 mice in each group and the animals were fed a diet adequate in vitamin E, which contained 0.05 mg selenium/kg.

Sodium selenite added to the drinking-water at a level of 9 mg selenium/litre killed all 12 rats fed a diet based on either ground corn or ground rye in 6 weeks (Palmer & Olson, 1974), whereas Halverson et al. (1966) found that sodium selenite added to a diet based on ground wheat, at a level of 9.6 mg/kg, killed only 1 out of 10 rats in the same period of time. However, the rats used in the latter study were post-weaning animals weighing 60 - 70 g when the study started, whereas the rats used in the former study were 21-day-old weanlings and weighed only 35 - 45 g. Other research workers have shown that the resistance of rats to the toxic effects of selenium increases markedly after the twenty-first day of life (Franke & Potter, 1936).

Feng et al. (1985) studied the hepatoxicity of high selenium corn (7.12 mg/kg) produced in the Enshi county of

China, where human selenium intoxication was reported (section 8.1.1.1). After feeding a diet that contained 61% of this corn (total selenium concentration of 4.343 mg/kg) for 16 weeks, liver cirrhosis was seen in 3 out of 6 male and 5 out of 6 female rats in one group. No liver damage was found histologically in another group of rats in the same study, which had consumed a diet containing 30.5% of the high selenium corn (total selenium concentration of 2.35 mg/kg).

At present, the best indicator of chronic selenium toxicity appears to be growth inhibition (US NAS/NRC, 1976), and a selenium level of 4 - 5 mg/kg is necessary to achieve this response in animals fed a normal diet. In laboratory rats, this exposure to selenium represents an intake of about 200 - 250 µg/kg body weight per day. More sensitive and specific criteria of selenium poisoning to demonstrate effects at lower dose levels, such as biochemical or histological techniques, would obviously be highly desirable, but such tests are not available at present.

7.1.2.3 Inhalation toxicity

The effects of respiratory exposure to selenium compounds, administered under conditions mimicking occupational exposure, have been described in several papers. Filatova (1951) studied the toxic effects of respiratory exposure to selenium dioxide (SeO_2) under conditions similar to those that occur in industry, i.e., heating of selenium (section 5.2.1). In acute studies, white rats were exposed to air concentrations of selenium dioxide of 0.15 - 0.6 mg/litre, and all rats died within one-half - 4 h. Morphological examination of the organs revealed that intra-alveolar and perivascular oedema occurred in the lungs, and haemorrhages and degenerative changes in the liver, kidney, and heart. In 4 additional studies, all rats survived 4 h when exposed to doses of 0.09, 0.06 - 0.07, or 0.03 - 0.04 mg selenium dioxide/litre, but all rats exposed to the highest dose (equal to 5 - 5.2 mg/kg body weight) died within 24 h. In a series of long-term studies, rats were exposed to repeated doses of selenium dioxide at 0.01 - 0.03, 0.006 - 0.009, or 0.003 - 0.005 mg/litre for 6 h, every other day, for one month. The lowest dose did not produce any effects on body weight or on the blood picture, and all the rats survived. Histological examination revealed degenerative changes in the liver, renal tubules, dystrophy of heart muscle, and hyperaemia and hypertrophy of the splenic pulp. At the dose of 0.006 - 0.009 mg/litre, all the rats but one died within 27 - 33 days. For the first 2 weeks, there was no difference in body weight between the exposed and unexposed control rats but, during the last 2 weeks, the exposed rats lost body weight and all but one of the exposed rats died. The histopathological changes consisted of multiple necrosis and degeneration in the liver and myocardial fibres, and involvement of renal tubules. In the third group of rats, which was exposed to 0.01 - 0.03 mg/litre, the animals showed respiratory distress, weight loss, and, in 3 rats, anaemia. All the rats died between days 8 and 18 of exposure. In the liver, kidneys, myocardium, and spleen, the changes observed were similar to those seen at lower doses, but more pronounced. Moreover, lung oedema similar to that noted in the acute exposure studies was seen.

Lipinskij (1962) exposed 2 groups of 5 rabbits to airborne amorphous selenium and selenium dioxide in chambers, under conditions analogous to industrial exposure, except that the doses were higher. In the first group, the rabbits were exposed to 20 µg selenium dioxide/litre and 40 mg selenium/m^3, for 2 h per day, for one week, at which time a decrease in blood catalase (EC 1.11.1.6) activity was noted. The rabbits in the second group were exposed to 10 µg selenium dioxide/litre and 20 mg selenium/m^3, for 2 h daily, for 12 weeks. After 12 weeks, decreases in total and reduced glutathione were observed, but there was no change in levels of oxidized glutathione.

On the basis of studies on 60 white rats weighing 120 - 150 g, Burchanov et al. (1969) concluded that intratracheal injection of 0.06 ml of a sterile suspension containing 50 mg of highly dispersed elemental selenium dust in physiological saline, for 1 - 12 months, resulted in decreases in body weight and muscular strength, and morphological and biochemical alterations in the respiratory tract. Burchanov (1972) obtained similar results in inhalation studies on rats exposed to 2 types of polymetallic dusts found in industry.

The acute toxicity of selenium dust (average mass median particle diameter, 1.2 µ) for rats, guinea-pigs, and rabbits was described by Hall et al. (1951). Exposure of these animals, for 16 h, to an atmosphere containing approximately 30 mg selenium dust/m^3 produced mild interstitial pneumonitis in the animals. Rats exposed to selenium fumes developed acute toxic effects, and it was suggested that the fumes might have contained some selenium dioxide.

The acute toxic effects of hydrogen selenide were investigated in guinea-pigs exposed for 10, 30, or 60 min to concentrations ranging from 0.002 to 0.57 mg/litre (Dudley & Miller, 1941). All animals exposed to 0.02 mg/litre, for 60 min, died within 25 days; 93% of those exposed to 0.043 mg/litre, for 30 min, died within 30 days, and all exposed to 0.57 mg/litre, for 10 min, died within 5 days. Decreasing the concentrations of hydrogen selenide, and increasing the time of exposure to 2, 4, or 8 h produced death in 50% of the guinea-pigs, within 8 h.

Acute studies on the toxicity of selenium hexafluoride (SeF_6), were carried out on the rabbit, guinea-pig, rat, and mouse at 100, 50, 25, 10, 5, and 1 ppm for 4 h. Exposures down to and including 10 ppm (Ct = 40 ppm/h) were uniformly fatal (Kimmerle, 1960). Exposure to 5 ppm (Ct = 20 ppm/h) resulted in pulmonary oedema from which the animals recovered, and 1 ppm did not induce any grossly observable effects. However, with repeated exposure for 1 h daily at 5 ppm, for 5 days, there were definite signs of pulmonary injury.

7.1.3 Blood levels in toxicity

Analyses of blood from cattle have shown that, if the average value of blood-selenium for a herd is over 2 mg/litre, damage from chronic selenium poisoning is very likely to occur (Dinkel et al., 1957). Average values of 1 - 2 mg/litre suggest a borderline problem, especially in reproduction, whereas values below 1 mg/litre indicate that no damage from toxicity should be expected. Rosenfeld & Beath (1964) commented that typical concentrations of selenium in the blood in alkali disease, blind staggers, and acute selenium poisoning were 1 - 2, 1.5 - 4, and up to 25 mg/litre, respectively. In studies with sodium selenite, Maag et al. (1960) found that severe selenium toxicity occurred in cattle when the selenium content of the blood exceeded 3 mg/litre. The selenium levels in blood associated with chronic toxicity in sheep corresponded to 0.6 - 0.7 mg/litre (Ermakov & Kovalskij, 1974).

7.1.4 Effects on reproduction

Franke & Potter (1936) fed 7 groups of 6 weanling rats (2 males and 4 females), 21 days of age, a basal diet consisting of ground whole wheat, 82%; commercial casein, 10%; pure leaf lard, 3%; dehydrated yeast, 2%; cod liver oil, 2%; and McCollum's salt mixture No. 185, 1%. Group 1 was shifted immediately to a toxic diet in which sufficient control wheat was substituted by seleniferous wheat to give a final selenium concentration of 24.6 mg/kg diet. Groups 2, 3, 4, 5, and 6 were shifted to the toxic diet when the rats attained 42, 63, 84, 105, and 186 days of age, respectively. Group 7 was fed the basal diet throughout the entire study. No matings were successful in which both males and females had been fed the toxic diet for 40 days, regardless of the age at which the rats had been taken off the basal ration. This was true, even in group 6, in which successful matings had been achieved at 82 and 131 days, while the rats were still being fed the basal diet. All male rats not placed on the toxic diet until 63 days of age or more were able to fertilize the normal females.

In a later study in which females fed the toxic diet were mated with normal males, there were some successful matings, but the pups that were born generally died or were eaten by the mother soon after birth.

In a study by Munsell et al. (1936), 7 groups of 4-week-old rats, each containing 2 males and 6 females, were fed a basal diet consisting of wheat, 58%; skim milk powder, 30%; yeast, 5%; butter, 5%; and cod liver oil, 2%. In each of the groups, a variable proportion of the control wheat was replaced by toxic wheat so that the latter contributed 8.7, 6.0, 3.0, 1.5, 0.75, 0.38, or 0 mg selenium/kg diet. Compared with the controls receiving no toxic wheat, the number of females having young was decreased by feeding the diets containing 3.0 mg or more selenium/kg, and the percentage of young reared was decreased by the diets containing 6.0 mg or more/kg (Table 25). Feeding the diet containing 0.75 mg/kg appeared to improve reproductive performance in terms of total young produced or percentage of young reared. In the second generation of rats continued on their respective levels of toxic grain, the diet containing 6.0 mg/kg had a definite deleterious effect on reproduction, whereas diets containing 0.75 - 1.5 mg/kg had some beneficial effects. In a second study, the percentage of young reared was improved in the group fed the diet containing 1.5 mg/kg compared with the control diet group, but the average number of young per litter and the average number of young per bearing female were decreased in the group fed the diet containing 0.75 mg/kg.

Halverson (1974) fed 4 groups of 70-day-old ARS/Sprague Dawley albino rats a basal diet containing glucose, 61.8%; purified casein, 25%; corn oil, 5%; minerals, 6%; and vitamins, 2.2%. The groups comprised 2 - 4 males and 6 - 8 females and were given the basal diet supplemented with sodium selenite at 0, 1.25, 2.50, or 3.75 mg selenium/kg. After 90 days, the rats were mated; no consistent effects of selenium on reproduction were observed in these first generation rats. Four groups of second-generation young were maintained on their respective diets and used as breeding stock in a second life cycle. In 2 separate studies, the first and second reproductions of these second generation rats were successful, regardless of the selenium content of the diet (Table 26). In the third reproduction, however, the unsupplemented rats showed signs of reproductive failure. Selenium supplementation at all levels prevented poor litter production in the first study but only levels of 2.50 and 3.75 mg/kg prevented it in the second. Neonatal survival was improved in both studies by selenium administered at 3.75 mg/kg.

Schroeder & Mitchener (1971a) gave selenate at 3 mg selenium/litre, in the drinking-water, continuously, during a multigeneration study in mice. Five pairs of F_0 mice were

Table 25. Effect of seleniferous wheat on reproduction in first- and second-generation rats[a]

Level of dietary selenium (from wheat) (mg/kg)	Generation and number of females 1st generation	Generation and number of females 2nd generation	Females having young Number	Females having young %	Total young Number	Total litters Number	Average number young/litter Number	Average number young/bearing female Number	Young reared Number	Young reared %
0	6	–	6	100	70	13	5.4	11.7	30	42.9
.38	6	–	5	83.3	72	13	5.5	14.4	23	31.9
.75	6	–	6	100	90	18	5.0	15.0	75	83.3
1.5	6	–	5	83.3	42	10	4.2	8.4	15	35.7
3.0	6	–	3	50.0	38	7	5.4	12.7	19	50.0
6.0	6	–	4	66.7	30	8	3.8	7.5	5	16.7
8.7	6	–	2	33.3	7	2	3.5	3.5	0	0
0	–	11	10	90.9	134	19	7.1	13.4	46	34.3
.38	–	9	6	66.7	63	10	6.3	10.5	31	49.2
.75	–	41	29	70.7	434	52	8.3	15.0	328	75.6
1.5	–	7	7	100	100	11	9.1	14.3	72	72.0
3.0	–	7	5	71.4	61	9	6.8	12.2	25	41.0
6.0	–	2	2	100	17	3	5.7	8.5	0	0
8.7	–	0	–	–	–	–	–	–	–	–

a Adapted from: Munsell et al. (1936).

Table 26. Effect of selenium as selenite on successive reproductions of second-generation rats maintained on a casein diet[a]

Study	Added selenium in diet	Number of brood females	Number of litters born			Number young per newborn litter			2-day survival of litter members (%)		
			1	2	3	1	2	3	1	2	3
	(mg/kg)										
1	0	3	3	3	0	12	12	–	61	46	–
	1.250	6	6	6	5	12	7	8	29	21	26
	2.500	5	5	5	5	12	11	8	71	84	30
	3.750	6	6	6	6	11	9	8	79	78	78
2	0	7	7	6	2	8	10	6	74	82	33
	1.250	7	7	5	2	10	11	4	90	84	00
	2.500	9	8	8	6	9	9	5	81	85	25
	3.750	8	7	7	4	7	6	6	92	100	92

[a] Adapted from: Halverson (1974).

given selenium at weaning and allowed to breed ad libitum for 6 months. Controls were treated identically but did not receive any selenium in the water. At weaning, F_1 pairs were randomly selected from the first, second, and third litters and left to breed, to produce the F_2 generation. F_2 pairs were similarly selected from the first or second litters to produce the F_3 generation. An increased male:female ratio was observed in all generations (1.30 – 1.50 versus 0.94 – 1.03 for the controls), but the mechanism of this effect was not explained. No breeding failures were seen in the control group, but 2 or 3 such failures occurred in each of the F_1, F_2, and F_3 generations given selenate. Only 3 litters were produced in the F_3 generation treated with selenate and 16 of 23 mice born were runts, whereas in the control F_3 generation, 22 litters were produced with a total of 230 mice and no runts.

Two groups of 5 male and 2 groups of 5 female Wistar rats were fed a laboratory chow diet and one group of each sex was given sodium selenate at 0 or 7.5 mg selenium/litre in the drinking-water, respectively (Rosenfeld & Beath, 1954). The rats receiving selenium ("selenized rats") were treated from birth to 8 months of age. Mating of normal males with selenized females was totally unsuccessful, whereas mating of selenized males with normal females resulted in normal reproduction and survival. Doses of 1.5 or 2.5 mg/litre did not have any effect on the reproduction of breeding rats, the number of young reared by the mothers, or the reproduction of 2 successive generations of males and females. However, a level of 2.5 mg/litre decreased the number of young reared by the mothers by about 50%.

Poley & Moxon (1938) fed 4 groups of 15 Rhode Island pullets a basal laying ration consisting of ground corn, 25%; ground barley, 25%; ground wheat, 15%; wheat bran, 8%; wheat middlings, 8%; meat and bone scraps, 8%; alfalfa leaf meal, 5%; dried buttermilk, 5%; salt, 0.5%; and cod liver oil concentrate, 0.5%. Each of the 4 groups was mated with 2 male brothers for the entire study. After 2 weeks, grains containing selenium (corn, barley, and wheat) were substituted for normal grains in order to contribute 0, 2.5, 5, or 10 mg selenium/kg diet fed to each of the 4 groups, respectively. Oyster shells and water were provided ad libitum. There were no significant differences in feed consumption, body weight, egg production, or egg fertility in any of the groups. After 4 weeks on the toxic diets, 10 mg of selenium/kg reduced hatchability to zero. Hatchability was only slightly reduced by 5 mg/kg, and a level of 2.5 mg/kg had no effect.

In a study by Wahlstrom & Olson (1959), 2 groups of 10 purebred Duroc gilts, 8 weeks of age, were fed a basal diet supplemented with sodium selenite at 10 mg selenium/kg diet.

Of the 9 gilts in the unsupplemented group (one was killed accidentally), 8 conceived with the first service and one failed to conceive after 3 services. Of the 10 gilts in the selenium-exposed group, 5 conceived with the first service, 2 with the second service and 3 failed to conceive after 3 services. The selenium-exposed group farrowed fewer litters and fewer live pigs, and weaned fewer pigs than the control group (Table 27). Moreover, the average birth weights and weaning weights were lower in the selenium-exposed group. The average litter weight at 56 days for both litters was 131 and 227 pounds for the selenium-exposed and unexposed groups, respectively.

Table 27. The effect of selenium on reproduction and lactation in swine

Items	Number litters farrowed	Average number pigs farrowed	Average number pigs farrowed alive	Average birth weight (kg)	Average number pigs weaned	Average weaning weight (kg)
Basal						
First litters	7	8.0	8.0	2.95	7.3	30.5
Second litters	6	11.2	10.5	2.89	8.0	29.0
Average	6.5	9.6	9.25	2.92	7.65	29.75
Basal + selenium						
First litters	6	9.8	7.7	2.60[b]	5.7	23.1[c]
Second litters	5	8.6	7.2	2.76	5.6	23.5[c]
Average	5.5	9.2	7.45	2.68	5.65	23.3

[a] Adapted from: Wahlstrom & Olson (1959).
[b] Significantly less than basal lot ($P < .05$)
[c] Significantly less than basal lot ($P < .01$)

Dinkel et al. (1963) reported on the effect of early (May 1 to mid-July) versus late (mid-July to October 1) breeding on the reproductive performance of beef cattle grazing on a seleniferous range in South Dakota. With such a programme, it was felt that early conception would avoid any deleterious effects on reproduction of the highly seleniferous young range grasses that grow luxuriantly during the early summer. Data from 152 matings, 75 in the early and 77 in the late breeding group, collected over 5 years, showed that the average conception rate of the early-bred cows was 60%, whereas that of the late-bred cows was 37.7%. Moreover, the average calf crop weaned was 52% in the early breeding group and 32.4% in

the late group. The data do not prove, but suggest, that
selenium was the cause of the difference. In addition to the
early breeding effect, the very low overall reproductive rate
should be noted. This is considerably below the rate that can
be attained with similar operations when chronic selenosis is
not a problem, and it has been repeatedly observed that, on
ranches with a selenium problem, the rate of reproduction is
consistently low. This led Olson (1969) to comment that the
effect on reproduction might be the most significant effect of
excessive amounts of the element from an economic standpoint.

7.1.5   Effects on dental caries

Most studies on experimental animals have shown that high
doses of selenium do not have any effect on caries formation
when the selenium is given after tooth formation has already
occurred. For example, Muhler & Shafer (1957) fed 22 male
weanling Sprague Dawley rats a stock corn cariogenic diet that
contained 10 mg sodium selenite/kg. After 6, 8, and 12 weeks,
the level of sodium selenite was raised to 15, 20, and 30
mg/kg diet, respectively. A control group of 20 rats received
the same diet without added selenium. The animals were given
their respective diets and drinking-water low in fluorine
(F = 0.2 mg/litre) ad libitum for 140 days. The selenium-
exposed group grew only about 53% as much as the controls, but
the mean number of carious lesions was essentially the same in
both groups (6.4 and 6.9, respectively). Claycomb et al.
(1965) fed 2 groups of 24 weanling rats a high-carbohydrate
diet, with or without 10 mg sodium selenite/kg diet, for 100
days. In this study, selenium depressed growth by only 11%,
but again the average number of carious lesions per rat was
similar in the selenium-treated and control groups (0.92 and
1.08, respectively). The rather low incidence of dental
caries was thought to be due to genetic influences. Wheatcroft
et al. (1951) maintained 80 white rats, 40 males and 40
females, on a coarse corn cariogenic diet for 100 days. The
rats were divided into 4 groups of 20 and were given daily
intraperitoneal doses of sodium selenite at 0, 0.2, 0.5, or
1.0 mg selenium/kg body weight. The authors felt that there
was a trend towards an increased incidence of caries in the
group receiving the most selenium, but the 2 highest doses of
selenium were clearly toxic, since these rats had diarrhoea,
affected eyes, and dull, matted fur. There was also histological evidence of liver damage in over half, and kidney
damage in a third, of the rats. Thirty new-born Wistar rats
were divided into 3 groups receiving, intraperitoneally, 0,
0.5 - 1.0 µg, or 5 - 10 µg sodium selenate, respectively,
every day for 53 days (Kaqueler et al., 1977). The group
receiving the lower dose of selenium had a lower total number

of caries than the controls, while the group receiving the
higher dose had more caries than the controls. The authors
concluded that post-eruptive administration of sodium selenate
may have either a cariostatic or cariogenic influence in rats,
depending on the dosage administered.

Attempts to induce dental caries by exposure to high
levels of selenium were more successful when the selenium was
given during the time of tooth development. For example,
Navia et al. (1968) fed sodium selenite at 4 mg selenium/kg,
to groups of 21 CR-COBS rats in the drinking-water, in a
purified caries-producing diet, or in the same diet gelled
with equal parts of 2% aqueous agar. The experimental
treatments of the mothers and litters began at birth and
continued until the pups were 50 days old. Selenium had no
effect on buccal, sulcal, or proximal caries, when given in
the diet, but caused a 12% increase in sulcal lesions when
administered in the drinking-water. Buttner (1963) fed 3
groups of 8 female, caries-susceptible rats of the Wistar
strain a cariogenic coarse stock corn diet plus 0, 5, or 10 mg
sodium selenite/litre drinking-water during mating, pregnancy,
and lactation. The offspring received the same concentrations
of sodium selenite for 120 days. Thirty-one pups were born in
the control group but, since the dosages of selenium used
partially inhibited reproduction, only 20 and 7 pups were born
in the groups receiving 5 or 10 mg sodium selenite/litre
drinking-water, respectively. Growth of the pups was strongly
inhibited and the number and extent of carious lesions were
increased in both selenium-treated groups (Table 28). In a
study by Bowen (1972), 7 female monkeys of the species Macaca
irus were fed a cariogenic diet, with drinking-water sweetened
with 3% sucrose, at night, and phosphate-free icing sugar, 4
times daily. A second group of 3 females of the same species
was treated identically except that they received selenium, as
sodium selenate, at 2 mg/litre drinking-water. After 15
months, some of the selenium-treated monkeys experienced
slight gastrointestinal trouble in the form of greenish
malodorous stools and the dose of selenium was reduced to
1 mg/litre for another 45 months. The second permanent molars
and the premolars were formed during the period of higher
selenium exposure and these teeth had a yellow chalky
appearance in the selenium-treated monkeys. On the other
hand, selenium had no effect on the first permanent molars,
which had already formed before the start of the study.
Although both groups of monkeys had severe dental caries, the
mean caries score of the selenium-treated group was about
twice that of the controls (16 and 8.4, respectively). There
was no difference between the times required for caries to
develop in the first permanent molars in the controls and
experimental groups, but carious lesions in the second

Table 28. Effect of developmental and postdevelopmental administration of sodium selenite on dental caries in the rat[a]

Dose of sodium selenite in drinking-water (mg/litre)	Sex	Number of rats	Weight gain in 120 days (g)	Mean number of carious lesions	Extent of carious lesions
0	M	17	330 ± 9	5.3 ± 0.5	11.6 ± 1.4
	F	14	207 ± 7		
5	M	13	240 ± 5	7.2 ± 0.7	18.7 ± 2.2
	F	7	176 ± 7		
10	M	5	220 ± 18	8.6 ± 1.1	25.2 ± 3.2
	F	2	148 ± 3		

[a] Adapted from: Buttner (1963).

permanent molars developed more rapidly in the selenium-exposed group (6.7 versus 21 months for mean caries development time in selenium-treated compared with control groups). The author concluded that selenium had a cariogenic effect, when administered during tooth development and a moderate anti-cariogenic effect, when given posteruptively.

Britton et al. (1980), however, presented evidence that showed that, under some conditions, selenium can have a cariostatic effect, even when given during tooth development. These workers gave 0, 0.8, or 2.4 mg selenium/litre drinking-water in the form of selenomethionine or sodium selenite, to rats from the 10th day of pregnancy until the pups were weaned. At 17 - 19 days of age, the pups were given oral innoculations of Streptococcus mutans - 6715 in thioglycollate broth. When 19 days old, the pups were weaned, divided into groups of 13 - 19, and given a 67% sucrose diet and distilled water ad libitum. At 65 - 68 days of age, the young rats were killed and the first and second molar teeth were stained with murexide and scored for caries. The high dose of selenium given to the mothers, as either compound, did not have any effect on dental caries, but the middle dose (0.8 mg/litre) had a definite cariostatic effect, since selenomethionine and sodium selenite decreased the incidence of total buccal caries by 46.1 and 40.9%, respectively. The authors suggested that this cariostatic effect might occur because some selenium is necessary for proper enamal formation or because of undetermined effects on the oral environment.

The fact that high doses of selenium have an apparent cariogenic effect, only when given during tooth development, is consistent with the results of Shearer (1975) who found that the incorporation into teeth of radioactive selenite or selenomethionine given in the drinking-water at a level of 0.2 mg selenium/litre was much greater in rat pups undergoing dental development than in their mothers whose teeth had already matured. The mechanism of the cariogenic effect of high levels of selenium is not known, but an antagonism to fluoride does not seem to be involved (Shearer & Ridlington, 1976).

7.1.6   Factors influencing toxicity

7.1.6.1 Form of selenium

Studies comparing the toxicity of different forms of selenium are described in section 7.1.2.1. This point is particularly relevant because the forms of selenium in foods have yet to be characterized.

7.1.6.2 Nutritional factors

High levels of dietary protein protect against selenium toxicity (Gortner, 1940), but some proteins give better protection than others (Smith & Stohlman, 1940). Jaffe (1976) found that the toxicity of seleniferous sesame meal for rats was markedly decreased when the diets were supplemented with L-lysine, the limiting essential amino acid in sesame proteins. Lysine was thought to protect against selenium poisoning indirectly by improving the biological value of the sesame proteins.

Linseed meal contains a unique non-proteinaceous factor that is highly effective in protecting against selenium poisoning (Halverson et al., 1955; Levander et al., 1970). The results of recent work suggest that the protective activity of linseed meal may reside in 2 newly-isolated cyanogenic glycosides (Palmer et al., 1980; Smith et al., 1980). Cyanide has a partially protective effect against selenium poisoning in rats (Palmer & Olson, 1979) and increases the occurrence of nutritional myopathy in lambs (Rudert & Lewis, 1978).

There is conflicting evidence as to the ability of methyl-donating compounds to protect against selenium toxicity (Rosenfeld & Beath, 1964). Methionine was shown to protect against chronic selenite poisoning in rats, but only when vitamin E or other fat-soluble antioxidants were in the diet (Levander & Morris, 1970).

Vitamin E deficiency increased the susceptibility of rats to chronic poisoning by selenium as selenite (Witting &

Horwitt, 1964) and pigs deficient in vitamin E and selenium
were more susceptible to acute selenium toxicosis than
non-deficient pigs (Van Vleet et al., 1974).

7.1.6.3 Arsenic

Selenium poisoning caused by feeding rats seleniferous
wheat was decreased by giving sodium arsenite in the drinking-
water (Moxon, 1938). The protective effect of arsenic against
selenium toxicity might be explained by the increased biliary
excretion of selenium caused by arsenic (Levander & Baumann,
1966). However, the use of arsenic compounds did not prove to
be a practical way of controlling selenium poisoning in livestock (Olson, 1969). Also, arsenic does not protect against
all forms of selenium, since it potentiates the toxicity of
the trimethylselenonium ion (Obermeyer et al., 1971).

7.1.6.4 Sulfate

Dietary sulfate partially counteracts the toxicity of
selenate in rats, but has little or no effect against selenite
or organic forms of selenium (Halverson et al., 1962). Sulfate
apparently increases the urinary excretion of selenium fed as
selenate, thereby reducing the retention of selenium in the
internal organs (Ganther & Baumann, 1962).

7.1.6.5 Adaptation

Ermakov & Kovalskij (1968) fed 18 adult female sheep from
seleniferous and normal localities their own respective diets
or these diets plus an additional 2 mg selenium, as selenite,
daily for 28 days. The 18 sheep were divided into 6 equal
groups, according to their history of selenium intake and, in
the case of the seleniferous sheep, according to the presence
or absence of signs of selenium toxicity. After 28 days, all
the sheep were killed, and selenium levels and acid and alkaline phosphatase activities were determined in several
tissues. Selenium levels and alkaline phosphatase activities
are presented in Table 29. The increase in selenium retention
in the blood, kidneys, and most of the other internal organs,
produced by supplementing the feed with selenite, was less in
sheep from high-selenium localities than in animals from
control localities. Retention in the liver seemed to be an
exception. The increase in alkaline phosphatase activity due
to the selenite load was much clearer in sheep from the
control localities than in the sheep from the seleniferous
localities. The alkaline phosphatase activity of the pancreas
in sheep from the high-selenium locality, manifesting selenium
toxicity signs, was an exception being much higher after the

Table 29. Effect of high selenium (Se) load on selenium metabolism and alkaline phosphatase activity in sheep previously exposed to normal and high selenium intakes in forage[a]

History of sheep	Daily Se intake			Blood Se (μg/litre)	Skeletal muscle Se (μg/kg)	Liver Se (μg/kg)	Kidney Se (μg/kg)	Alkaline phosphatase activity (units/100 g tissue)		
	Plant Se (μg)	Selenite Se (μg)	Total Se (μg)					kidney	liver	skeletal muscle
High Se-curved hoofs	1821	2000	3821	411±18	230±24	600±44	870±69	286±148	448±43	13.0±2.7
High Se-curved hoofs	1944	0	1944	290±18	180±7	330±20	760±26	362±89	374±117	9.5±3.0
High Se-normal hoofs	1774	2000	3774	362±34	210±16	590±90	1000±174	309±103	335±18	9.4±2.9
High Se-normal hoofs	2038	0	2038	239±34	170±6	340±20	890±21	349±63	344±56	9.5±1.2
Normal Se	376	2000	2376	408±23	170±8	450±38	950±77	496±53	457±57	14.8±3.0
Normal Se	396	0	396	100±12	89±5	190±20	780±69	143±29	254±7	5.5±0.8

[a] Adapted from: Ermakov & Kovalskij (1968).

selenite load (370 ± 57 compared with 84 ± 34). These differences in the response of the sheep to selenite loading were interpreted by the authors as evidence of the adaptation of the animals to high levels of selenium exposure.

Jaffe & Mondragon (1969) presented evidence that animals could adapt to long-term selenium intake. Hepatic-selenium levels decreased steadily in young rats fed a diet containing 4.5 mg selenium/kg as seleniferous sesame meal, if the rats came from mothers that had been fed the same seleniferous diet, but the levels increased if the rats came from mothers fed a non-seleniferous stock diet. It was concluded that an adaptation mechanism existed that allowed rats exposed to long-term selenium ingestion to store less of this element than previously unexposed controls. A later report (Jaffe & Mondragon, 1975) confirmed these results and also indicated that the hepatic lesions, splenomegaly, and other toxicity signs were significantly decreased in adapted rats.

One mechanism for a possible adaptive response to high levels of selenium could involve effects of selenium intake on various selenium metabolic pathways, since previous selenium exposure has been shown to have an influence on the whole body retention of selenium and the urinary and pulmonary excretion of selenium (Ganther et al., 1966; Ewan et al., 1967; Burk et al., 1973) (section 6.3, 6.4). However, as discussed in section 7.1.2.1, more recent work has shown that previous selenium exposure directly influences the toxicity of final selenium metabolites such as dimethyl selenide and the trimethylselenonium ion (Parizek et al., 1976, 1980).

7.1.7   Mechanism of toxicity

The mechanism of selenium toxicity remains unclear and the modes of action of various selenium compounds, such as hydrogen selenide, selenomethionine, and selenium dioxide are likely to be quite different. Thus, no unifying hypothesis regarding the toxic effects of selenium compounds is possible. Various hypotheses concerning the mechanisms of selenium toxicity have been presented (Rosenfeld & Beath, 1964; Izraelson et al., 1973; Ermakov & Kovalskij, 1974; US NAS/NRC, 1976; Lazarev, 1977; Levander, 1982). Although it appears unlikely that selenite interferes directly with sulfhydryl enzymes (Tsen & Tappel, 1958; Tsen & Collier, 1959), it is possible that selenite may interfere with glutathione metabolism (Vernie et al., 1978; Chung & Maines, 1981; Vernie, 1984; Annudi et al., 1984) and that this may affect enzyme activities. The apparent pro-oxidant nature of high levels of selenite has been reported by several workers (Witting & Horwitt, 1964; Csallany et al., 1984), and this may be related to some aspects of selenite toxicity.

7.2   Selenium Deficiency

7.2.1   Animal diseases

The nutritionally beneficial effects of selenium were first reported in 1957 by Schwarz, who discovered that sodium selenite prevented dietary liver necrosis in vitamin E-deficient rats (Schwarz & Foltz, 1957). This discovery led to the rapid recognition of vitamin E-related selenium-responsive deficiency diseases in several species of farm animals. Diseases such as white muscle disease in sheep and cattle, hepatosis dietetica in swine, and exudative diathesis in poultry are economically significant problems in areas of the world where the soil levels of selenium available for uptake by plants are low. Schwarz & Foltz (1958) also reported that feeding a diet deficient in both selenium and vitamin E to mice resulted in multiple necrotic degeneration of several organs; more recently, Suchkov et al. (1977, 1978) found that feeding such diets to mice for 13 days caused a decreased staining intensity for zinc in the pancreas, kidneys, and testes.

The fact that vitamin E as well as selenium tended to protect against many of these diseases led some research workers to question whether there was any requirement for selenium in animals receiving adequate amounts of vitamin E. But, the results of more recent work have clearly demonstrated the need for selenium, even in animals given nutritionally adequate levels of vitamin E. Deficiency signs specific for selenium included alopecia, vascular changes, cataract, poor growth, and reproductive failure in second generation selenium-deficient rats (McCoy & Weswig, 1969; Wu et al., 1979). Pancreatic degeneration occurred in selenium-deficient chicks with a normal intake of vitamin E (Thompson & Scott, 1970), but the condition could be prevented by feeding very high levels (> 300 µg/kg) of vitamin E or other antioxidants (Whiteacre et al., 1983).

Changes in the myocardial parenchyma and increases in heart weight have been observed in albino rats fed corn and vegetables grown in the areas in which Keshan disease is endemic (Su et al., 1982), and swine fed grain from endemic areas for 6 months exhibited multiple myocardial necrosis as well as other lesions (Zhu et al., 1981). Vitamin E deficiency exacerbated the pathological and histochemical changes in the heart and liver of piglets fed a low selenium diet composed mainly of cereals grown in the Keshan disease area (Zhu et al., 1981).

7.2.2   Intakes needed to prevent deficiency

7.2.2.1   Quantitative dietary levels

Schwarz (1961) showed that 0.02 - 0.03 mg selenium/kg diet, in the form of sodium selenate or selenite, afforded 50% protection against liver necrosis in vitamin E-deficient rats over an experimental period of 30 days. Selenite-selenium at 0.10 mg/kg protected the rats over their entire life span in the absence of vitamin E. Under thse conditions, the rats developed severe tocopherol deficiency, which mainly afflicted the central nervous system. Sprinker et al. (1971) noted that 0.10 mg selenium/kg, as selenite, prevented specific selenium deficiency lesions in rats that had been fed a vitamin E-supplemented diet for 2 generations. Thompson & Scott (1969) found that administration of 0.02 - 0.05 mg selenium/kg, as selenite, prevented death and exudative diathesis in chicks fed diets containing typical levels of vitamin E (sections 7.2.4.1 and 7.2.7.2 include discussions on the nutritional interrelationship of selenium and vitamin E).

Considerable field experience in New Zealand has indicated that feeds from pastures associated with selenium-responsive unthriftiness in sheep contain 0.008 - 0.030 mg selenium/kg (Hartley, 1967). In areas where selenium-responsive diseases do not occur, the feed levels of selenium range from 0.020 - 0.098 mg/kg. Since there is some overlap in these selenium levels, other factors may be involved in the etiology of this selenium-responsive disease (various factors that can influence selenium deficiency are discussed in section 7.2.7).

It has been concluded that "the critical level for dietary selenium, below which deficiency symptoms are observed, is apparently about 0.02 mg/kg for ruminants and 0.03 - 0.05 mg/kg for poultry" (US NAS/NRC, 1971). However, this US NAS/NRC committee pointed out that "when supplementary selenium is fed, higher levels than the minimal requirements have been proposed, both to permit satisfactory distribution of the element through the large feed mass and to overcome variations in feed intake by individual animals". The final recommendation of the committee was that 0.1 and 0.2 mg selenium should be added per kg feed to eliminate selenium deficiency in livestock and turkeys, respectively. In laboratory rats, a dietary-selenium level of 0.1 - 0.2 mg/kg would be equivalent to an intake of about 5 - 10 µg/kg body weight per day.

7.2.2.2 Bioavailability

7.2.3 Blood and tissue levels in deficiency

Hartley (1967) presented data on the blood-selenium levels
of normal New Zealand lambs and lambs suffering from various
selenium-responsive diseases (Table 30). The mean blood-
selenium levels were lower in the diseased lambs than in the
healthy ones, but there was considerable overlap between the
two groups. It was suggested that other unknown factors might
be involved in the etiology of these diseases. Jacobsson et
al. (1970) found blood-selenium levels of 0.003 mg/litre in
cows afflicted with muscular degeneration, whereas healthy
cows had an average of 0.046 mg/litre. Hartley (1967)
estimated that blood-selenium levels of 0.05 mg/litre in young
sheep could be considered satisfactory.

Table 30. Blood-selenium levels in normal and selenium-deficient lambs[a]

Group	Selenium content[b] (mg/litre)
Normal[c]	0.026 (0.014 - 0.048)
White muscle disease[d]	0.016 (0.006 - 0.033)
Normal[e]	0.06 (0.020 - 0.195)
Selenium-responsive unthriftiness[f]	0.01 (0.007 - 0.030)

[a] Adapted from: Hartley (1967).
[b] Mean with range in parentheses.
[c] Normal lambs from areas with white muscle disease.
[d] Lambs with white muscle disease.
[e] Lambs from areas where selenium-responsive diseases are not
     recognized.
[f] Lambs with selenium-responsive unthriftiness.

The same authors compared liver-selenium levels found in
normal lambs with those found in lambs with selenium-
responsive unthriftiness (Table 31). Again, there was some
overlap between the two groups, but the average level of
selenium in the livers from the deficient lambs was much lower
than that from the healthy lambs. Allaway et al. (1966)
suggested that 0.21 mg/kg (dry weight) is the critical minimal
hepatic-selenium level below which a high incidence of white
muscle disease can be expected in lambs.

Table 31. Liver-selenium levels in normal lambs or
lambs with selenium-responsive unthriftiness[a]

Group	Selenium content[b] (mg/kg)
Normal[c]	0.16 (0.03 - 0.40)
Selenium-responsive unthriftiness[d]	0.018 (0.005 - 0.05)

[a] Adapted from: Hartley (1967).
[b] Mean with range in parentheses.
[c] Lambs from areas where selenium-responsive diseases are not recognized.
[d] Lambs with selenium-responsive unthriftiness.

7.2.4 Physiological role: glutathione peroxidase

7.2.4.1 Function of selenium and relationship to vitamin E

The discovery that selenium is a component of the enzyme glutathione peroxidase (Rotruck et al., 1973) provides a logical explanation for the nutritional interaction of vitamin E and selenium, a puzzle that perplexed scientists for many years. Analysis of purified ovine erythrocyte glutathione peroxidase by fluorometry revealed that the protein contained 4 moles of selenium per mole of enzyme (Hoekstra et al., 1973) and this stoichiometry was also found for crystalline bovine erythrocyte-glutathione peroxidase analysed for selenium by neutron activation analysis (Flohe et al., 1973). In the scheme postulated by Hoekstra (1975a), tocopherols act as intracellular antioxidants to prevent oxidative damage to polyunsaturated fatty acids in biological membranes by terminating chain reactions of lipid peroxides (Fig. 7). Selenium, as a part of glutathione peroxidase, protects against oxidative stress either by catalysing the destruction of hydrogen peroxide or by catalysing the decomposition of lipid hydroperoxides, thereby interrupting the free radical peroxidative chain reaction; in this latter role the free fatty acyl hydroperoxide must be liberated by phospholipan A_2 action from the hydrophobic region of biological membranes, before its reduction can be catalysed by glutathione peroxidase (Grossman & Wendel, 1983). Thus, both of these nutrients play separate but interrelated roles in the cellular defence mechanisms against oxidative damage.

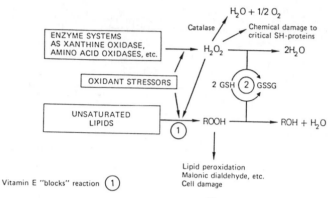

Vitamin E "blocks" reaction (1)

Se, as a component of GSH-Peroxidase, catalyses reaction (2)

WHO 85912

Fig. 7. Physiological role of selenium and its relationship to the
antioxidant function of vitamin E. Modified from: Hoekstra
(1975a).
(1) Vitamin E terminates chain reactions of lipid peroxidation.
(2) Se, as a component of GSH-peroxidase, catalyses reaction.

Although glutathione peroxidase accounts for many of the
biological effects of selenium, other functions of this trace
element in the body may yet be discovered. Only 1/5 of the
total selenium in rat brain is in the form of glutathione
peroxidase (Prohaska & Ganther, 1976). Selenium has also been
shown to be a constituent of several enzymes in microorganisms
(Stadtman, 1977).

7.2.4.2 Effect of selenium intake on tissue-glutathione
peroxidase activity

The results of several studies have shown a close dose-
effect relationship between the dietary intake of selenium and
the activity of glutathione peroxidase in various tissues. For
example, Hafeman et al. (1974) found that increased dietary
selenium caused corresponding increases in erythrocyte-glutathione peroxidase activity in rats, even when toxic levels of
selenium were fed (Fig. 8). Hepatic-glutathione peroxidase
activity increased with increasing dietary-selenium levels up
to 1 mg/kg, but then declined due to liver damage when toxic
levels were fed (Fig. 9). However, studies on hamsters have
shown that consumption of excess selenium does not always
result in increased erythrocyte-glutathione peroxidase activity, since feeding selenite at more than 10 mg selenium/kg
diet did not cause increases in erythrocyte-enzyme activity

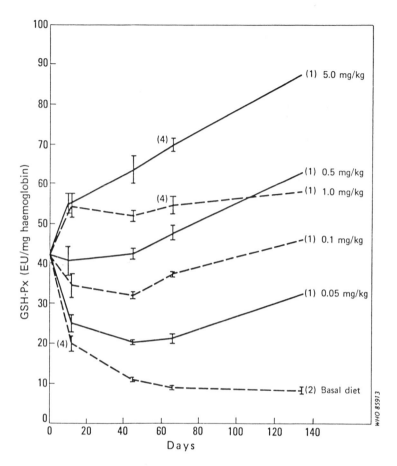

Fig. 8. Relationship between dietary-selenium intake and units of glutathione peroxidase activity (EU) in rat erythrocytes. From: Hafeman et al. (1974).

that corresponded to increases in blood-selenium concentrations (Julius et al., 1980).

Oh et al. (1976a) investigated the effects of feeding several different levels of dietary selenium on the activity of glutathione peroxidase in various tissues in lambs. The activity of the enzyme tended to plateau in all tissues examined, except red cells and pancreas, when 0.1 mg was fed per kg diet. This level of dietary selenium approximates the presumed requirement in this species.

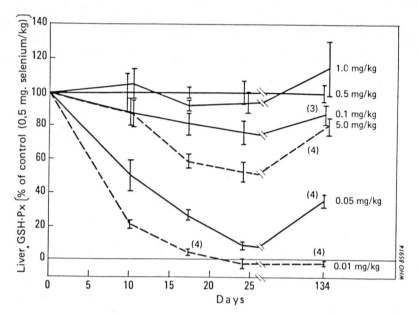

Fig. 9. Relationship between dietary-selenium intake and glutathione peroxidase activity in rat liver. From: Hafeman et al. (1974).

7.2.4.3 Relationship between blood-selenium levels and erythrocyte-glutathione peroxidase activity

A direct linear relationship between the selenium concentration of whole blood and the activity of erythrocyte-glutathione peroxidase has been demonstrated in lambs, sheep, cattle, and swine (Oh et al., 1976b; Wilson & Judson, 1976; Sivertsen et al., 1977). This linear relationship in lambs (Fig. 10) was thought to be due to the fact that glutathione peroxidase accounts for most of the total selenium in the ovine red cell and that ovine red cells contain a relatively constant proportion of the total blood-selenium.

Under certain conditions, some investigators have not found a significant correlation between blood-selenium levels and blood glutathione peroxidase activity. For example, Thompson et al. (1976) described swine that had high blood-selenium levels but low blood-glutathione peroxidase activities. The reasons for this discrepancy are not known but could perhaps be related to an increased level of "non-functional" selenium in the blood, such as selenomethionine that has been nonspecifically incorporated into protein as a substitute for methionine.

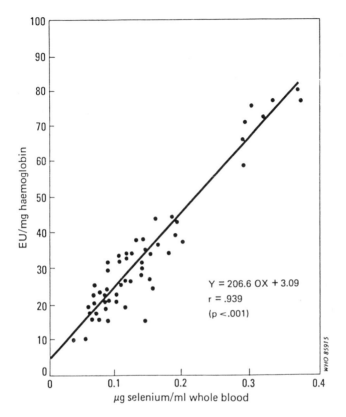

Fig. 10. Relationship between blood-selenium level and units of erythrocyte-glutathione peroxidase activity (EU) in lambs. From: Oh et al. (1976b).

7.2.4.4 Glutathione peroxidase as an indicator of selenium status

As discussed by Hoekstra (1975b), the measurement of glutathione peroxidase activity offers several advantages as an indicator of selenium status. First of all, the enzyme assay is easier to perform and is less time consuming than the selenium analysis. Also, the enzyme assay is not subject to contamination problems. Moreover, glutathione peroxidase represents only "functional" selenium in the tissues, not selenium that has been non-specifically incorporated into proteins or that has formed biologically inactive complexes with heavy metals. Finally, at least in some tissues, the activity of the enzyme appears to taper off as the dietary level of selenium approaches the nutritional requirement of the animal. This "plateau effect" of glutathione peroxidase activity does not occur in red blood cells, a biological material that is highly convenient for sampling purposes. However, as pointed out by Hoekstra (1975b), the absence of plateauing in red cells may limit the usefulness of erythrocyte-glutathione peroxidase in accurately establishing selenium requirements, but does not diminish the usefulness of this source of the enzyme as an overall index of selenium status.

One complicating factor in the use of glutathione peroxidase to assess selenium status is the fact that the activity of the enzyme is influenced by many physiological variables other than selenium intake (Ganther et al., 1976). Among these are the age and sex of the animal, starvation, exposure to certain oxidant stressors, toxicants, or heavy metals, and deficiency in iron and vitamin B_{12}. Obviously, these variables have to be controlled, or compensated for, if glutathione peroxidase is to be a valid index of selenium status.

Another serious complicating factor in the use of glutathione peroxidase to assess selenium status is the existence of a "non selenium-dependent" glutathione peroxidase activity that persists, even in severe selenium deficiency (Lawrence & Burk, 1976). This enzymic activity has been found in variable amounts in different rat tissues ranging from 23% of the total glutathione peroxidase activity in fat to 91% in testis (Lawrence & Burk, 1978). The "non selenium-dependent" activity accounts for 26 - 38% of the total glutathione peroxidase activity in rat brain, kidneys, liver, and adrenals. No such activity is found in rat erythrocytes, spleen, heart, thymus, intestinal mucosum, skin, or skeletal muscle. The "non selenium-dependent" glutathione peroxidase activity has been found in the liver of several different species including the rat, hamster, guinea-pig, chicken, pig, sheep, and man. In human and guinea-pig liver, it accounts for the major part of the total glutathione peroxidase activity. The "non selenium-dependent" glutathione peroxidase activity differs from the selenium-containing enzyme in substrate specificity in that it catalyses the reduction of a range of organic hydroperoxides but not of hydrogen peroxide. The activity of the "non selenium-dependent" enzyme is inversely related in rat liver to the dietary level of selenium (Lawrence & Burk, 1978; Lawrence et al., 1978) and the glutathione-S-transferase enzymes are responsible for the "non selenium-dependent" enzyme activity (Prohaska & Ganther, 1977; Lawrence et al., 1978). The glutathione-S-transferases are not selenium-containing enzymes; they are all dimeric, and their monomeric constituents fall into three categories of different relative molecular mass. The glutathione peroxidase activity of the enzymes is only shown by the dimeric enzymes that contain one of the three subunits and, in selenium deficiency, the amount of this subunit alone is increased (Ketterer et al., 1982; Mehlert & Diplock, 1985).

The physiological significance of the "non selenium-dependent" activity has been questioned because the seleno-enzyme has a much higher V_{max} and lower K_M (Prohaska & Ganther, 1977). However, Burk et al. (1980a) showed that the "non selenium-dependent enzyme" removed organic hydroperoxides in intact rat livers perfused with a haemoglobin-free medium containing high concentrations of peroxides.

Obviously, any use of glutathione peroxidase activity to monitor selenium status has to take into account this "non selenium-dependent" activity. This complication can be avoided by selecting a tissue for test that has little or none of the selenium-independent activity, such as erythrocytes, or by using hydrogen peroxide as the substrate, since this has little activity with the selenium-independent enzyme.

7.2.5   Other possible physiological roles

7.2.5.1   Homeostasis of hepatic haem

Burk et al. (1974) showed that the induction of hepatic cytochrome P-450 by daily intraperitoneal injections of 75 mg phenobarbital/kg body weight, for 4 days, was markedly impaired in rats that had been fed a selenium-deficient diet for at least 3 months. The induction of ethylmorphine demethylase activity by phenobarbital was also impaired in the livers of selenium-deficient rats. However, no such impairment was observed in the induction of cytochrome b_5 or of NADPH-cytochrome c reductase activity or biphenyl hydroxylase activity, or in pentobarbital sleeping time (Burk & Masters, 1975). Treatment with phenobarbital stimulated a 6- to 8-fold increase in hepatic microsomal haem oxygenase activity in selenium-deficient rats but had little or no effect in selenium-supplemented rats (Correia & Burk, 1976). This suggested that the abnormalities of cytochrome P-450 induction in selenium-deficient rats were related to increased degradation rather than decreased synthesis of hepatic haem. Mackinnon & Simon (1976) reported an impaired haem synthesis in selenium-deficient phenobarbital-treated rats, but Burk & Correia (1977) were unable to confirm these results. Apparently, the pronounced stimulation of microsomal haem oxygenase (EC 1.14.99.3) in the livers of selenium-deficient rats by phenobarbital is unrelated to the role of selenium in gluta-thione peroxidase, since pretreatment of deficient rats with a single dose of selenite at 50 mg selenium/rat, 4 - 6 h before phenobarbital administration, abolished the response in haem oxygenase but had no effect on cytosolic or mitochondrial glutathione peroxidase activity (Correia & Burk, 1978). Moreover, phenobarbital failed to stimulate liver microsomal haem oxygenase activity or to affect cytochrome P-450 in vitamin E-deficient, selenium-adequate rats, so this response does not seem to be related to the direct effects of lipid peroxidation on haem or cytochrome P-450. There is a marked sex difference in the phenobarbital stimulation of hepatic microsomal haem oxygenase activity in selenium-deficient rats. The stimulation was greatest in males and least in females with intermediate values in castrated males and testosterone-treated females (Burk et al., 1980a). This sex difference existed even though hepatic glutathione peroxidase activity was reduced to the same extent by selenium deficiency in all groups. However, the exact mechanism by which selenium influences homeostasis of hepatic haem still needs to be clarified. Pascoe et al. (1983) showed that both intraluminal iron and selenium were required by the intestinal mucosa for maintenance of the intestinal cytochrome P-450 content, and its dependent mixed function oxidase activity, and that withdrawal of dietary selenium for one day markedly decreased the level of cyto-chrome P-450 in the small intestine, suggesting that intes-tinal mucosal cells derive their selenium from the diet rather than from the blood.

A report by Maines & Kappas (1976) indicated that injec-tion of selenium increased hepatic haem oxygenase activity, but the doses of selenium used (3.94 mg/kg body weight) exceeded the MLD for selenium as selenite (Table 19), so that the phenomenon observed by these authors may have been merely an artifact of acute selenium poisoning.

7.2.5.2 Microsomal and mitochondrial electron transport

Diplock and co-workers carried out an extensive series of studies, the results of which suggest a possible role for selenium and vitamin E in the electron-transfer system of rat-liver microsomes (Diplock, 1974a,b). First, they showed that appreciable amounts of volatile selenium evolved from rat liver subcellular organelles, after acid treatment, only when the rats were adequate in vitamin E and the tissue homo-genization medium contained large concentrations of anti-oxidants (Diplock et al., 1971). The acid-volatile selenium was later identified as selenide (Diplock et al., 1973). It was assumed that this compound was formed in vivo, presumably by the glutathione reductase pathway (Fig. 9), but artifactual in vitro formation of such selenide, because of the presence of mercaptoethanol in the homogenization medium, cannot be ruled out (Levander, 1976a). An oxidant-labile form of non-haem iron, dependent on dietary vitamin E and selenium, was observed in rat liver microsomes (Caygill & Diplock, 1973). It was considered that this supported the hypothesis that the selenide might form part of the active centre of a non-haem iron protein functional in microsomal electron transfer. Alterations in the kinetic parameters of aminopyrine demethylation by the liver microsomes in vitamin E-deficient rats also supported this idea (Giasuddin et al., 1975). However, attempts to isolate an oxidant-labile selenium- and vitamin E-dependent non-haem, iron-containing protein fraction were unsuccessful, because of the lability of the selenide, and it was concluded that clarification of the role of selenide in microsomes needed additional research (Diplock, 1976, 1979). The effects of prolonged selenium deficiency on a large number of parameters in mouse liver xenobiotic metabolism were studied by Reiter & Wendel (1983). They demonstrated a heterogeneous pattern of effects that did not appear to involve glutathione peroxidase.

One of the earliest biochemical defects discovered in vitamin E- and selenium-deficient rats was the inability of liver slices or homogenates to maintain normal rates of respiration after prolonged periods of incubation in vitro (Schwarz, 1962). Although this so-called respiratory decline might be explained on the basis of generalized mitochondrial membrane damage due to lipid peroxidation (Yeh & Johnson, 1973), it was also plausible that selenium might have a more specific role in mitochondrial metabolism. Indeed, selenium added to the diet, or added in vitro, accelerated the swelling of rat liver mitochondria induced by certain thiols, and this swelling could be blocked by the addition of cyanide (Levander et al., 1973a). Selenium was in fact shown to be an excellent catalyst for the reduction of cytochrome c by thiols in chemically-defined systems (Levander et al., 1973b) and a selenoprotein containing a haem group similar to that of cytochrome c has been reported in lamb muscle (Whanger et al., 1973). However, more work is required to establish the possible role of selenium in mitochondrial electron transfer.

7.2.5.3 The immune response

Dietary vitamin E in excess of nutritionally required levels has been shown to improve the humoral immune response of several different species to bacterial and viral antigens (Nockels, 1979). Because of the close nutritional and biochemical relationship between selenium and vitamin E (section 7.2.4.1), the effect of dietary selenium supplementation was tested on the immunological responses of mice (Spallholz et al., 1973). Mice fed a laboratory chow diet (thought to contain 0.5 mg selenium/kg) supplemented with sodium selenite at 0.7 or 2.8 mg selenium/kg had approximately 7- and 30-fold higher antibody titres, respectively, after challenge with sheep red blood cells than mice fed the chow diet alone. Sodium selenite injected into mice intraperitoneally at doses of 3 - 5 µg per animal also increased the primary immune response to the sheep red-blood-cell antigen and this increase was greatest when the selenium was given prior to, or simultaneously with, the antigen (Spallholz et al., 1975). The mechanism by which selenium stimulates antibody synthesis is not known (Martin & Spallholz, 1976), but the dose of selenium needed to elicit this response is clearly above the nutritionally required level. Levander (1986) has reviewed the very limited data on the possible effect of selenium on the immune response in man.

7.2.5.4 Selenium and vision

The effects of selenium on light perception were studied in 50 male "Grey chinchilla" rabbits weighing 3 - 3.5 kg and maintained on a standard laboratory diet (Abdullaev et al., 1972, 1974a,b). Electroretinography was carried out in dark-adapted animals under conditions of maximally dilated pupil induced by topical application of 0.5% dikaine - 1% atropine. Light stimulation was provided with an impulse lamp giving either single or dual flashes of 150 microseconds. Eight different energy levels of light intensity were used ranging from 0.016 to 1.4 J. Sodium selenite was administered subcutaneously at a dose of 0.45 mg selenium/kg body weight. Control rabbits were injected with sodium sulfite at a dose of 1 mg/kg. Sodium selenite stimulated light sensitivity, as judged by increases in the a and b waves of the electroretinogram (ERG). The increase in the amplitude of the b wave was considerably greater than that of the a wave. This stimulating effect of sodium selenite on light sensitivity persisted for 3 days in all studies. The same effects were observed when sodium selenite was given via retrobulbar administration. Sodium sulfite did not have any effect on the ERG. Similar results were obtained by Bacharev et al. (1975).

The Task Group recognized that the data on the effects of selenium on vision were obtained in studies involving high dose levels (in the case of rabbits, the doses were approximately 20 - 30% of the LD_{50}). The effects on vision of lower doses of selenium have not been elucidated so far. The possible effect of deficiency of selenium on vision is also of interest. Model studies with isolated retinas exposed to a 0.01% concentration of sodium selenite resulted in an increase in the amplitude of the ERG (Kulieva et al., 1978). The effect of increased sensitivity of the eye to light, after administration of sodium selenite, and the rate of manifestation and duration of the effect are directly correlated with localization of the substance in the structures of the eye (section 6.2.2.1). Subcutaneous administration of sodium selenite to rabbits, at a dose of 0.4 - 2 mg/kg body weight, inhibited free radical formation as indicated by electron paramagnetic resonance (EPR) measurements, not only in the lipid membrane phase but also within the melanoprotein granules of the retinal epithelium. The EPR signal was decreased in intensity by selenite treatment in comparison with controls, but the form and parameters of the EPR lines were not changed (Abdullaev et al., 1974a,b).

7.2.6   Effects on reproduction

Vitamin E was discovered as a result of the isolation of an unknown fat-soluble dietary factor essential for reproduction in rats (Evans & Bishop, 1922), but selenium compounds are not active in preventing reproductive failure in vitamin E-deficient rats (Harris et al., 1958). Moreover, female rats weighing 80 - 120 g and fed a selenium-deficient, but vitamin E-adequate, diet successfully delivered 3 litters over a period of 6 months, even though their blood-selenium levels dropped from 0.52 to 0.06 mg/litre during this time (Hurt et al., 1971). Similarly, McCoy & Weswig (1969) found that rats fed a low-selenium but adequate-vitamin E diet, for 4 months, reproduced normally. However, their offspring failed to reproduce if kept on the low selenium diet. Immobile spermatozoa, with separation of heads from tails, were observed in 5 out of the 8 male progeny not given selenium supplements, and no spermatozoa were seen in the other 3 males. The 2 male and 2 female offspring, supplemented with sodium selenite at 0.1 mg selenium/kg diet, were fertile but delivered litters of only one or 2 rats, all of which died in a few days. The authors concluded that selenium supplementation resulted in some fertilization, but only a few full-term young were delivered and these were abnormal as they survived for only a brief period.

Wu et al. (1973) found that the motility of spermatozoa from male rats, born to females fed a selenium-deficient diet, was always very poor and that most of the sperm cells showed breakage near the midpiece of the tail. Supplementation of the diet with high levels of vitamin E or various other antioxidants did not prevent these selenium deficiency effects. However, in this study, the body and testicular weights of the rats were so markedly depressed by selenium deficiency that it was not possible to establish whether the abnormal sperm were due to a direct effect of selenium deficiency or to an indirect effect on overall growth and well-being. In a second series of studies (Wu et al., 1979), using male rats born to dams fed a diet adequate in selenium, the depression of growth, and testes weight in the progeny caused by selenium deficiency was much less, even though impaired sperm motility and abnormal sperm morphology were still observed (Table 32). The authors also pointed out that there did not appear to be any correlation between reduction in body or testicular weights and the characteristic sperm midpiece abnormality among the individual rats in each treatment group. However, all of the rats used by Wu and associates were fed diets that were probably low in zinc and chromium and both of these trace elements are known to play important roles in sperm formation (Wu et al., 1971; US NAS/NRC, 1979; Anderson & Polansky, 1981).

The mechanism of the effects of selenium on the integrity of sperm morphology in the rat is not known, but Brown & Burk (1973) found over 40% of an injected tracer dose of radio-selenite in the testis-epididymis complex of selenium-deficient rats, after 3 weeks. Autoradiography of epididymal sperm revealed that the labelled selenium was heavily localized in the midpiece. Calvin (1978) has isolated a selenopolypeptide, with a relative molecular mass of 17 000 daltons, from rat sperm, which may have a critical function in the normal assembly of the sperm tail.

Selenium cannot prevent resorption-gestation in vitamin E-deficient rats, nor can it improve reproductive performance in vitamin E-deficient chickens or turkeys (Creger et al., 1960; Jensen & McGinnis, 1960). However, studies in which severe depletion of selenium was produced by the long-term feeding of low-selenium, high-vitamin E diets have shown the beneficial effects of the element on egg production and hatchability in poultry. For example, Cantor & Scott (1974) fed 1-day-old Single Comb White Leghorn chicks semi-purified low-selenium starter, grower, and developer diets, all of which contained only 0.02 mg naturally-occurring selenium/kg from 0 to 6, 7 to 12, and 13 to 25 weeks of age, respectively. Because of the low selenium level, high levels of vitamin E (50 - 84 IU/kg) were added to the diets to prevent exudative diathesis during the growth period. From 25 weeks of age to 10 months, the birds were fed semi-purified low-selenium layer diets that contained 11 IU of vitamin E/kg. After 10 months, the hens were fed a low-selenium corn-soybean meal practical laying ration containing 0.027 mg natural selenium/kg, by analysis. The calculated vitamin E content was 16 IU/kg. At 12 1/2 months of age, 4 replicate groups of 5 hens were continued on the low-selenium diet, whereas a similar number of hens was fed the same diet supplemented with sodium selenite at 0.1 mg selenium/kg. Selenium supplementation had beneficial effects on both egg production and hatchability of fertile eggs (Table 33), but there was no significant effect of selenium on egg fertility, determined after 2 - 5 weeks (data not shown).

Table 32. Effects of selenium deficiency on body and testicular weights and spermatozoan morphology and motility in rats[a]

Dietary selenite supplement (mg selenium/kg)	Number of rats	Duration (month)	Body weight (g)	Testes weight (g)	Sperm motility[b]	Number of rats with sperm midpiece abnormality
Study 1						
0	6	4	452 ± 6	3.29 ± 0.06	6.2	0/6
0.5	6	4	470 ± 18	3.55 ± 0.11	6.2	0/6
0	4[c]	12	476 ± 17	3.03 ± 0.30	2.5	2/4
0.5	4[c]	12	494 ± 12	3.57 ± 0.07	5.8	0/4
Study 2						
0	10[c]	11	456 ± 16	3.07 ± 0.09	3.5	5/10
0.4	12	11	543 ± 18	3.50 ± 0.08	8.0	0/12

[a] Adapted from: Wu et al. (1979).
[b] Relative activity of sperm observed with a light microscope at 400x at 37 °C and graded from 0 - 10 with 10 representing the highest motility.
[c] Two rats died in each of these groups.

Table 33. Effect of dietary selenium supplements on reproductive
performance of 12 1/2-month-old selenium-depleted hens fed a low-
selenium, corn-soybean-meal practical ration[a]

Selenium supplement	None	0.1 mg selenium as Na_2SeO_3/kg diet
Criterion		
Egg production (%)		
Weeks 1-2	67.2	76.6
3-4	66.4	78.5
5-6	64.2	71.8
Day 46-76	54.2	79.7
77-107	55.1	76.7
108-137	59.1	75.1
Hatchability of fertile eggs (%)		
Week 2	57.3	93.0
3	54.7	97.1
4	54.5	94.9
5	42.4	86.6
17	0.0	90.3
18	8.0	93.6

[a] Adapted from: Cantor & Scott (1974).

Cantor et al. (1978) carried out a similar study on
turkeys in which large white breeder hens and toms were raised
from one day of age until sexual maturity on a series of
Torula yeast-containing diets, low in selenium content
(0.025 - 0.047 mg/kg). At 32 weeks of age, 8 replicate
groups, each comprising 6 individually-caged hens, were fed a
low-selenium basal diet (0.033 mg/kg) or the same diet
supplemented with sodium selenite at 0.2 mg selenium/kg.
Eight individually-caged toms were also assigned to each
dietary treatment at 32 weeks of age. Neither the tom nor the
hen dietary treatments had any effect on fertility and the tom
dietary treatment did not have any effects on hatchability.
But the hatchability of fertile eggs from hens fed the
unsupplemented and supplemented diets was about 50 and 59%,
respectively. The mechanism by which dietary selenium
improves hatchability in chickens or turkeys is not known, but
the improvement may be due to increased embryonic selenium
levels or to a reduced embryonic need of vitamin E (Cantor &
Scott, 1974).

In New Zealand, a close geographical relationship was
observed between the presence of white muscle disease and the

incidence of barren ewes (Table 34). A selenium intervention trial was carried out on 3 farms where there was a history of white muscle disease and low lambing percentages (Table 35). The selenium-treated ewes received 5 mg of selenium, as sodium selenite, at monthly intervals from one month before tupping until just before lambing; half of the rams were also given selenium at monthly intervals before and during tupping (Hartley et al., 1960; Hartley & Grant, 1961). Selenium-dosed groups on all farms had higher lambing percentages and fewer barren ewes than the controls. Moreover, white muscle disease was eliminated in the lambs of the treated ewes. Estrus, ovulation, fertilization, and early embryonic development all proceeded normally in selenium-responsive ovine infertility until between 3 and 4 weeks after conception (at about the time of implantation) when the embryos perished for unknown reasons (Hartley, 1963). The actual cause of selenium-responsive infertility is also unknown and the geographical distribution of the disease is not always the same as that of other selenium deficiency diseases, such as white muscle disease or selenium-responsive unthriftiness (Gardiner et al., 1962). This suggests either that selenium-responsive infertility occurs only where intake of selenium is particularly low or that some other factors must be present that exacerbate the effects of low selenium intake. For example, mercury has a strong antagonism with selenium (section 7.4.1), and Parizek et al. (1971) showed that inorganic mercury could block the transport of selenium from mother rats to the fetuses.

Table 34. Geographical relationship between congenital white muscle disease and incidence of barren ewes[a]

Congenital white muscle disease	Number of farms	Incidence of barren ewes		
		up to 10%	11 - 30%	over 30%
Present	21	3	8	10
Absent	16	12	4	0

[a] Adapted from: Hartley et al. (1960).

7.2.7 Factors influencing deficiency

7.2.7.1 Form of selenium

Schwarz (1961) pointed out that only 0.007 mg naturally-occurring selenium/kg diet in extracts from pork kidney

Table 35. Effects of selenium (Se) on the reproductive performance of
sheep and white muscle disease in lambs[a]

Farm	Lambing		Barren ewes		White muscle disease incidence	
	control	Se	control	Se	control	Se
	(%)		(%)		(%)	
C	60.6	94.9	31.1	7.5	37.5	0
D	55.0	86.0	24.3	11.5	22.2	0
E	70.5	93.6	26.1	6.4	12.0	0

[a] Adapted from: Hartley et al. (1960).

("Factor 3 Selenium") was needed to afford 50% protection
against dietary necrotic liver degeneration in rats, whereas
0.020 - 0.030 mg/kg was required when the selenium was in the
form of selenite, selenate, selenocystine, or selenomethionine. Elemental selenium was essentially inactive.
Many organic derivatives were synthesized in an attempt to
develop selenium compounds that were not highly toxic but
still retained considerable biological potency against
deficiency diseases (Schwarz et al., 1972). Simple amino acid
derivatives of monoseleno diacetic acid appeared promising in
this regard and the hope was expressed that some of these
compounds might make it possible to use selenium in the
prevention and cure of disease (Schwarz, 1976).

Cantor et al. (1975a) studied the ability of selenium in
various poultry feed ingredients to prevent exudative
diathesis, a selenium deficiency disease, in chickens. In
general, the selenium in plant products was more readily
available than that in animal products, but the latter
consisted of highly-processed fish or poultry meals.

Selenium in the form of selenomethionine was less
effective than selenium as sodium selenite for preventing
exudative diathesis in chicks (Noguchi et al., 1973), but the
reverse was true for the prevention of pancreatic fibrosis
(Cantor et al., 1975b). In fact, selenomethionine was four
times as effective as either selenite or selenocystine for
this purpose; this phenomenon may have been related to the
peculiar affinity of this compound for the pancreas.

7.2.7.2 Vitamin E

Vitamin E decreases the level of dietary selenium needed
to prevent deficiency diseases in several species of animals.
For example, Scott & Thompson (1971) demonstrated that chicks

receiving 100 mg vitamin E/kg diet needed only 0.01 mg
selenium/kg diet to prevent exudative diathesis, whereas
chicks not receiving any added vitamin E needed 0.05 mg/kg.
Similarly, Scott et al. (1967) showed that selenium at 0.18
mg/kg protected against gizzard myopathy in turkey poults fed
diets containing vitamin E, but that 0.28 mg/kg was needed to
protect poults fed diets not supplemented with vitamin E.
Hakkarainen et al. (1978) found that 0.135 mg selenium/kg was
insufficient to prevent deficiency signs in swine fed a diet
containing 1.5 mg vitamin E/kg but was adequate in swine fed
5.0 mg vitamin E/kg. Increasing the level of dietary vitamin
E decreased the severity of nutritional muscular dystrophy in
lambs fed diets containing selenium at 0.1 mg/kg (Ewan et al.,
1968). In lambs not receiving any supplementary dietary
selenium, lambs receiving low levels of vitamin E died of
dystrophy before their hepatic-selenium levels were depleted
to 0.21 mg/kg, the level considered by Allaway et al. (1966)
to be the critical concentration of selenium in liver for the
development of nutritional muscular dystrophy (section 7.2.3).
With the highest level of vitamin E supplementation, the
liver-selenium values were lower than the critical level
suggested by Allaway et al. (1966) and yet the lambs still
survived. These results suggest that the tissue-selenium
level is not the only factor involved, but that the levels of
both tocopherol and selenium determine the appearance of
nutritional muscular dystrophy. The effects of vitamin E in
reducing selenium requirements are readily explainable on the
basis of the close biochemical relationship between these two
nutrients discussed in section 7.2.4.1.

7.2.7.3 Heavy metals and other minerals

Since selenium protects against the toxicity of several
heavy metals (section 7.4), many scientists have examined the
effect of heavy metals on the induction of selenium
deficiency. Silver accelerated the development of liver
necrosis in rats (Whanger, 1976) and increased the severity of
selenium-vitamin E deficiency in swine (Van Vleet, 1976;
McDowell et al., 1978), presumably by decreasing the activity
of glutathione peroxidase. Inorganic mercury also decreased
glutathione peroxidase activity in the rat but did not have
any effect on the rate of development of liver necrosis
(Whanger, 1976). However, methylmercury accentuated the
development of selenium deficiency in pigs fed a low-selenium
diet (Froseth et al., 1974). In spite of the ability of
arsenic to enhance the biliary excretion of toxic levels of
selenium (section 7.1.6.3), several attempts to induce
selenium deficiency by feeding arsenic have been unsuccessful
(Levander, 1977; McDowell et al., 1978). On the other hand,

drenching pregnant and lactating ewes with potassium cyanide, which partially counteracts selenium toxicity in rats (section 7.1.6.2), increased the incidence of nutritional myopathy in their lambs (Rudert & Lewis, 1978). It has been suggested that the use of sulfate fertilizers may decrease selenium concentrations in forage plants (Pratley & McFarlane, 1974; Westermann & Robbins, 1974), and that this could lead to an increased incidence of selenium deficiency in farm animals (Schubert et al., 1961). Dietary sulfate had no effect on the incidence of white muscle disease but increased the number of lambs with degenerative lesions of the heart (Whanger et al., 1969). Neither cadmium nor tellurium had any effect on white muscle disease in lambs (Whanger et al., 1976b) or vitamin E-selenium deficiency in swine (McDowell et al., 1978).

7.2.7.4 Xenobiotics

The results of early work by Hove (1948) suggested that vitamin E could protect rats fed a low-protein diet from death due to carbon tetrachloride (CCl_4) poisoning. The establishment of a nutritional relationship between selenium and vitamin E stimulated Seward et al. (1966) to study deficiencies of these nutrients in relation to poisoning (Table 36). Mortality was higher in the group deficient in both nutrients than in controls, but selenium deficiency did not increase mortality. Other workers (Hafeman & Hoekstra, 1977a; Burk & Lane, 1979) have confirmed this result. The increased mortality due to CCl_4 in doubly-deficient animals could have been due simply to inhibition of eating. Fasting these animals overnight can precipitate dietary liver necrosis leading to death (Hafeman & Hoekstra, 1977b).

Paraquat is another xenobiotic that is thought to exert its toxic effect via increased lipid peroxidation (Bus et al., 1974). Selenium deficiency worsens lung injury in rats caused by paraquat (Omaye et al., 1978) and leads to liver injury that would not otherwise occur in mice exposed to this compound (Cagen & Gibson, 1977). Burk et al. (1980b) showed that selenium-deficient rats were much more susceptible to paraquat or diquat poisoning than controls (Table 37). The deficient rats also produced more ethane and had a higher serum-glutamic-pyruvic transaminase (SGPT) activity, when treated with paraquat or diquat, than the controls. The same workers found that injected selenium protected deficient rats against diquat toxicity before appreciable amounts of glutathione peroxidase activity appeared in the tissues. They suggested the existence of a selenium-dependent factor, apart from glutathione peroxidase which protects against lipid peroxidation.

Table 36. Protective effect of selenium and vitamin E against
carbon tetrachloride toxicity in rats fed a 10% soybean-protein diet[a]

Dietary supplement	Weight gain (g/8 weeks)	Survival 48 h after CCl$_4$ injection[b]
None	62 ± 5	3/10
Vitamin E[c]	77 ± 3	9/10
Selenium[d]	79 ± 6	8/10
Vitamin E and selenium	80 ± 5	10/10

[a] Adapted from: Seward et al. (1966).
[b] All rats injected intraperitoneally with a single dose at 2 ml/kg body weight.
[c] Added as a mixture of equal quantities of dl-α-tocopherol and dl-α-tocopheryl acetate at 200 mg/kg diet.
[d] Added at 0.1 mg selenium/kg diet, as sodium selenite.

Combs et al. (1975) noted a similarity between the oedema produced in chicks fed a practical diet containing polychlorinated biphenyls (PCBs) and that observed in chicks fed a diet deficient in selenium and vitamin E (exudative diathesis). When selenium-depleted hens were fed a diet not containing PCBs, selenite at 0.15 mg selenium/kg diet was sufficient to prevent exudative diathesis in the chicks, but was not sufficient to prevent the disease when the hens were fed a diet containing 10 mg PCBs/kg (Table 38). In another study, addition of 50 mg PCBs/kg diet increased the incidence of exudative diathesis in vitamin E-deficient chicks receiving selenite at 0.04 mg selenium/kg diet from less than 60% in the controls to almost 90% in the PCBs group, 14 days after hatching (Combs & Scott, 1975). Dietary PCBs depressed plasma-glutathione peroxidase activity and increased the amount of dietary selenium required to protect liver microsomal fractions from peroxidation in vitro. On this basis, the authors concluded that dietary PCBs potentiate vitamin E-selenium deficiency in chicks by interfering with the biological utilization of dietary selenium, though some interference with the absorption and retention of vitamin E was also seen.

Burk & Lane (1979) pointed out the complexity of nutritional toxicology experimentation involving selenium, vitamin E, and various drugs and other chemicals. After studying the effects of numerous xenobiotics on rats that were

Table 37. Effect of selenium deficiency in rats treated with paraquat or diquat[a],[b]

Diet group	Agent given	Dose (μmol/kg)	Ethane produced (pmol/100 g/h)	SGPT activity (mU/ml)	Survival time (min)
Selenium-deficient	paraquat	78	470 ± 93	212 ± 87	297 ± 74
Control	paraquat	78	18 ± 13	32 ± 26	survived 24 h
Control	paraquat	390	92 ± 5	45 ± 34	106 ± 30
Selenium-deficient	diquat	19.5	1940 ± 420	3490 ± 1940	150 ± 37
Control	diquat	230	56 ± 50	41 ± 5	80 ± 12

[a] Adapted from: Burk et al. (1980b).
[b] Values are mean ± SD of 4 or 5 values.

Table 38. Effect of PCBs fed to hens on the incidence of
exudative diathesis in progeny fed a sub-optimal level of selenium[a]

Additions to diets of hens[b]		Exudative diathesis in chicks[c] (%)	Weight gain of chicks (g/2 weeks)
PCBs[d] (mg/kg)	Selenium[e] (mg/kg)		
0	0	35	89
0	0.15	0	102
10	0	30	82
10	0.15	45	78

[a]   Adapted from: Combs et al. (1975).
[b]   Single Comb White Leghorn hens, 27 weeks old, that had been reared on low-selenium, semi-purified diets supplemented with high levels of vitamin E; hens were on dietary treatments shown in Table for an additional 5 weeks.
[c]   Determined at 2 weeks of age. Chicks were fed a selenium-deficient diet (less than 0.02 mg selenium/kg) supplemented with 0.06 mg selenium, as Na_2SeO_3/kg. Diet was essentially free of tocopherols.
[d]   Aroclor® 1254.
[e]   Added as Na_2SeO_3.

deficient in either vitamin E or selenium, they concluded that in vivo lipid peroxidation, as assessed by ethane production, was not necessarily correlated with liver necrosis. Moreover, it was concluded that selenium deficiency was not just a lack of glutathione peroxidase activity in the tissues, since this condition also elevates hepatic glutathione S-transferase activity. Thus, selenium deficiency may actually protect against the hepatotoxicity of certain compounds, (e.g., acetaminophen and iodipamide) by increasing their conjugation with glutathione.

7.2.7.5 Exercise stress

Ancedotal observations from New Zealand suggested that selenium might be of value against "tying-up" in thoroughbred racehorses (Hartley & Grant, 1961), a condition that occurs during training and is associated with exercise. Characteristics include muscle stiffness and tenderness and disinclination to move. Young & Keeler (1962) applied mechanical restraint to one foreleg of lambs born to ewes fed a diet known to produce nutritional muscular dystrophy and found that lesions either did not occur or occurred to a much lesser extent in muscles of the restrained limb than in muscles of the unrestrained limb. Brady et al. (1978) observed increased levels of malondialdehyde, a product of lipid peroxidation, in the erythrocytes of 6 horses, immediately after exercise, but were unable to reproduce this phenomenon in a second exercise trial with 8 horses. Sodium selenite, given for 4 weeks in the trace mineral salt at a level calculated to provide 0.15 mg supplementary selenium/kg diet, did not have any effect on the elevated plasma-enzyme activities observed in exercised horses. Furthermore, it did not have any effect on plasma-selenium levels or erythrocyte-glutathione peroxidase activity. Harthoorn & Young (1974) commented that the pathological picture in wild animals that have died following mechanical capture ("capture myopathy" or "overstraining syndrome") is indistinguishable from white muscle disease seen in cattle suffering from vitamin E deficiency, but pointed out that prophylactic and subsequent symptomatic treatment with vitamin E and selenium-containing preparations did not have any beneficial effects on captured animals. Brady et al. (1979) expected to induce death in rats that had been fed a selenium and vitamin E-deficient diet for 4 weeks, after exercising them to exhaustion by swimming, but had no success, in spite of markedly depressed tissue-glutathione peroxidase activity, decreased hepatic stores of fat-soluble antioxidants, and increased levels of thio-barbituric acid-reacting substances. The authors concluded that the rat might be an unsuitable model for an exercise stress-susceptible species.

7.3 Ratio Between Toxic and Sufficient Exposures

The minimum level of dietary selenium that causes overt signs of chronic selenium toxicity in most species of animals is of the order of 4 - 5 mg/kg (US NAS/NRC, 1976). The minimum level of dietary selenium needed to prevent selenium deficiency diseases in most species is in the range of 0.02 - 0.05 mg/kg (US NAS/NRC, 1971). Therefore, the ratio of toxic to deficient selenium exposures is about 100. This difference between the harmful and beneficial levels of selenium is similar to that between the harmful and beneficial levels of several nutritionally essential minerals. However, in the case of selenium, this ratio can be decreased by certain nutritional or environmental factors. For example, a deficiency in vitamin E decreases the tolerance of animals to selenium toxicity (section 7.1.6.2) but increases the nutritional need for the element (section 7.2.7.2). Also, inorganic mercury increases the toxicity of methylated selenium metabolites (section 7.4.1) but methylmercury potentiates selenium deficiency (section 7.2.7.3). Thus, caution may have to be exercised in some situations to avoid a diminution in the ratio between harmful and beneficial intakes of selenium.

7.4 Protection Against Heavy Metal Toxicity

7.4.1. Mercury

Parizek & Ostadalova (1967) first showed that selenium could protect against the toxicity of mercury in short-term acute toxicity studies. This led to the suggestion that one of the nutritional roles of selenium might be protection of the organism against traces of toxic metals that enter the body, even under normal conditions (Parizek et al., 1971). Nutritional levels of dietary selenium have been demonstrated to decrease the chronic toxicity of the highly hazardous methylmercury (Ganther et al., 1972) and the selenium in tuna may provide a built-in protective mechanism against the methylmercury in such fish (Ganther & Sunde, 1974). The possible practical significance of the metabolic interaction of selenium and mercury has been indicated by the strong correlations observed between the concentrations of mercury and selenium in the livers of marine mammals under normal environmental conditions (Koeman et al., 1975) and in human tissues following exposure to inorganic mercury (Kosta et al., 1975). The mechanism by which selenium protects against the toxicity of methylmercury is not known, but the fact that vitamin E and certain antioxidants also decrease methylmercury toxicity (Welsh & Soares, 1976; Welsh, 1979) has prompted the hypothesis that these compounds may diminish methylmercury toxicity by counteracting the damaging effects of the free radicals generated by its breakdown (Ganther, 1978).

Although inorganic selenium salts protect against the toxicity of inorganic or methylmercury under a wide variety of conditions, dimethylselenide has a strong synergistic toxic action with mercuric chloride (Parizek et al., 1971). The mechanism of this synergism is unknown.

7.4.2 Cadmium

Parenterally administered selenium protects against the toxicity of injected doses of cadmium in rats, apparently by diverting the cadmium from low relative molecular mass target proteins to other proteins of higher relative molecular mass (Parizek et al., 1971; Chen et al., 1975). However, selenium does not cause this diversion when given orally, so any mechanism by which selenium counteracts cadmium toxicity, under environmental conditions, is yet to be determined (Whanger, 1976). The cadmium/selenium antagonism may be important for human health in that selenium prevents the hypertension caused by long-term cadmium exposure in rats (Perry et al., 1976).

7.4.3   Other heavy metals

Selenium may interact with lead but the interaction appears to be much weaker than that with mercury or cadmium, and vitamin E seems to be more important than selenium in determining the effects of lead poisoning (Levander, 1979). The interaction of selenium and silver is of theoretical interest since glutathione peroxidase activity is depressed by silver (Wagner et al., 1975), but this interaction does not appear to have any practical significance. Large doses of selenate protect against thallium toxicity in rats (Rusiecki & Brzezinski, 1966), but this interaction has not been investigated in any detail.

7.5   Cytotoxicity, Mutagenicity, and Anti-mutagenicity

7.5.1   Cytotoxicity and mutagenicity

Walker & Ting (1967) treated the soil around barley seedlings with solutions of sodium selenate to determine the effect on crossing over. Hoagland's solution containing 0, 0.5, 1.6, or 5.0 mg selenium/litre was applied 10 times, at 2-day intervals, to each of 4 groups of plants, each treatment consisting of the application of a 250 ml aliquot of the appropriate solution per pot. On the day after treatment, the pots were irrigated with distilled water to prevent lethal accumulation of selenate. Genetic data indicated a significant effect of the selenate on the rate of crossing over, consisting of a progressive decline with increasing concentration.

Walker & Bradley (1969) reared larvae of Drosophila melanogaster on a semi-defined culture medium containing selenocystine at concentrations of 0, 2, 10, and 50 µM. The selenocystine had different effects on crossing over on various chromosomal segments and the authors concluded that these effects were mediated through a structural chromosomal protein with a role in the exchange process.

In studies by Nakamuro et al. (1976), various inorganic compounds of selenium exhibited clastogenic effects when added in vitro to cultures of human leukocytes (Table 39). Quadrivalent selenium compounds were generally more efficient than hexavalent compounds in inducing chromosome aberrations. Similarly, selenite was more potent than selenate in causing DNA damage to Bacillus subtilis that was subject to recombination repair. The same authors also found that 1.6×10^{-3} mol H_2SeO_3 or SeO_2 caused about a 30% inactivation of the tryptophan marker of B. subtilis transforming DNA, whereas similar concentrations of Na_2SeO_4, H_2SeO_4, or Na_2SeO_3 were without any significant effect.

Table 39. Chromosome aberrations in leukocytes exposed to selenium compounds[a]

Compound	Dose	Occurrence of particular chromosome aberrations[b]				Total aberrations
		chromatid gaps	chromatid breaks	iso-chromatid breaks	chromatid exchanges	
	$(\times 10^{-5}$ mol)	aberrations per 100 cells				
Na_2SeO_4	53	8.4	0.9	0.9	0.9	11.2
	26	9.3	0.8	0.8	0.8	11.9
	13	11.2	0	0	0	11.2
H_2SeO_4	53	21.0	14.5	3.2	1.6	40.3
	26	11.9	0	0	0	11.9
	13	9.2	0.8	0	0	10.0
Na_2SeO_3	26	54.0	14.5	1.6	1.6	71.8
	13	41.7	8.3	0.9	0.9	51.9
H_2SeO_3	26	113.2	41.2	22.1	5.8	182.4
	13	92.3	42.3	12.8	2.6	150.0
	6.5	15.2	6.3	5.4	0	26.8
SeO_2	26	52.1	22.5	7.0	1.4	83.1
	13	24.2	2.5	3.3	0	30.0
	6.5	15.5	3.6	4.8	0	20.9
None	-	7.6	0	0	0	7.6

a Adapted from: Nakamuro et al. (1976).
b Number of cells examined varied from 62 - 250.

10

Lo et al. (1978) showed that selenite caused chromosome aberrations and mitotic inhibition in cultured human fibroblasts and that incubation with a mouse liver S9 microsomal fraction increased this capacity of selenite about 5-fold at equimolar concentrations (Table 40). Selenite also induced DNA fragmentation and repair synthesis and decreased the clone-forming capacity of the fibroblasts. Incubation with the S9 fraction increased the last 2 effects of selenite but slightly decreased the extent of DNA fragmentation. The response to selenite was similar in cultured fibroblasts from normal persons and from DNA repair-deficient xeroderma pigmentosum patients. The ability of selenate to trigger DNA repair, inhibit the mitotic rate, or cause a lethal effect was about the same as that of selenite over concentrations varying from 7×10^{-6} to 2×10^{-3} mol. However, when both compounds were incubated with the S9 fraction, only the activity of the selenite was potentiated (Table 41).

Table 40. Effect of 1.5 h selenite treatement on chromosome aberrations and mitotic rate of human fibroblasts[a]

Sodium selenite for 1.5 h (mol)	Metaphase plates with chromosome aberrations		Mitotic rate	
	without S-9 (%)	with S-9 (%)	without S-9 (%)	with S-9 (%)
Control	1.2	0.8	9.0	8.9
2×10^{-5}	1.3	6.9	8.6	9.8
4×10^{-5}	1.6	7.4	8.9	9.6
8×10^{-5}	3.0	14.3	9.3	9.4
1×10^{-4}	3.2	-	8.2	0.1
2×10^{-4}	4.7	-	9.5	0.0
4×10^{-4}	6.1	-	6.4	0.0
8×10^{-4}	14.5	-	6.9	0.0
1×10^{-3}	14.2	-	7.6	0.0
3×10^{-3}	18.1	-	0.7	0.0

[a] Adapted from: Lo et al. (1978).

Another example of a tissue fraction influencing the cytotoxicity of selenite is provided by the work of Ray & associates (Ray & Altenburg, 1978; Ray et al., 1978). These workers first showed that sodium selenite tripled the frequency of sister-chromatid exchanges (SCEs) in lymphocytes cultured with whole human blood (Table 42). At the concentrations of selenite used, the selenite could only be

Table 41. Effect of selenite versus selenate on human fibroblasts
incubated with, or without, mouse liver S9 microsomal fraction[a]

Treatment	DNA repair (grains/nucleus)	Mitotic rate (%)	Clone forming capacity (%)
Selenite[b]	10.7	5.2	36
Selenite + S9	115.6	0.0	0
Selenate	7.0	4.8	42
Selenate + S9	9.1	3.9	32

[a] Adapted from: Lo et al. (1978).
[b] The concentrations used for the estimation of DNA repair, mitotic rate, and surviving colonies were 8×10^{-4}, 2×10^{-7}, and 2×10^{-4} mol, respectively. Selenite and selenate were applied at equimolar concentrations for 1.5 h.

present during the final 19 h of the 96-h culture incubation, otherwise cell death occurred, as measured by mitotic indices. Later, they found that 7.90×10^{-6} mol Na_2SeO_3 increased SCE frequency, only when the lymphocytes were cultured in the presence of whole blood or separated red cells. Lysis of the erythrocytes indicated that the lysate rather than the red cell ghosts contained the activity necessary for selenite to raise the lymphocyte SCE frequency. The authors suggested that exposure of selenite to the lysate possibly resulted in a chemical modification of the selenite that enabled it to induce SCEs. This chemical modification of the selenite probably involves reduction to glutathione selenopersulfide derivatives, since Ray (1984) found that reduced glutathione (10^{-3} - 10^{-4} mol) could substitute for erythrocytes in activating sodium selenite to its SCE-inducing form. Whiting et al. (1980) showed that glutathione markedly stimulated unscheduled DNA synthesis in cultured human skin fibroblasts and chromosome aberrations in Chinese hamster ovary cells caused by a variety of inorganic selenium compounds. These research workers suggested that reduced selenium compounds are the ultimate mutagens and that their formation depends on reduction by glutathione.

In a preliminary note, Lofroth & Ames (1978) stated that selenite did not give any indication of mutagenicity (<< 0.01 revertants/n mol) whereas selenate gave rise to base-pair substitutions with about 0.03 revertants/n mol in a Salmonella plate incorporation test using different histidine-requiring strains that are reverted to prototrophy

Table 42. Effect of sodium selenite on sister-chromatid exchange frequency and mitotic index in lymphocytes cultured with whole human blood[a]

Na_2SeO_3 added (mol)	Sister chromatid exchanges per cell[b]	Average mitotic index
Control	6.74	0.03
1.58×10^{-6}	7.17	0.03
3.95×10^{-6}	7.49	0.03
7.90×10^{-6}	20.71	0.02
1.19×10^{-5}	21.51	0.03

[a] Adapted from: Ray et al. (1978).
[b] Average of 4 studies; a total of 100 cells was scored for each Na_2SeO_3 concentration except 75 at the highest concentration.

by different mechanisms. On the other hand, Noda et al. (1979) found that both selenite and selenate were weak mutagens in a similar test, since they gave rise to base-pair substitution (0.2 and 0.05 revertants/n mol, respectively). Ray & Altenburg (1980) showed that sodium selenide and sodium selenite were more active than sodium selenate in their ability to induce SCEs in human whole-blood cultures. Sirianni & Huang (1983) found that sodium selenite was the most potent inducer of SCEs in Chinese hamster V79 cells when S9 mixture was present, whereas sodium selenide was the most effective inducer in the absence of S9. For sodium selenate, there was no increase in SCE rate compared with controls, regardless of whether S9 was present or absent. At present, it is difficult to provide a biochemical explanation for the contrasting cytogenic effects exhibited by the various selenium compounds, when studied in different in vitro test systems.

Norppa et al. (1980) investigated the chromosomal effects of sodium selenite when given in vivo. They found that supplementation of human neuronal ceroid lipofuscinosis patients with tablets furnishing sodium selenite at a dose of 0.025 mg selenium/kg body weight for 1 - 13.5 months did not have any detectable effects on chromosomal aberrations or SCEs in peripheral blood lymphocytes. Similarly, mice treated with a single dose of sodium selenite at 0.8 mg selenium/kg body weight did not show any rise in chromosomal abnormalities in bone-marrow cells or primary spermatocytes after 24 h. However, there was an increased number of SCEs and chromosomal aberrations in the bone-marrow cells of Chinese hamsters, 17 - 19 h after injection with sodium selenite at a dose of 0.3 - 6.0 mg selenium/kg body weight. It was thought that the manifestations of chromosomal damage observed in the second study may have been related to the general toxicity of selenium at the high doses used.

7.5.2 Anti-mutagenicity

Shamberger et al. (1973a) tested the ability of sodium selenite and various antioxidants to decrease the chromosomal breakage induced by 7,12-dimethylbenz[a]anthracene (DMBA) in human blood leukocyte cultures. They found that 0.20 µmol sodium selenite reduced the breaks caused by 1.6 µmol DMBA by 42% (Table 43). This concentration of selenite was used because 1 µmol almost completely inhibited the growth of the cultures. In later studies, Shamberger et al. (1979) found that sodium selenite at concentrations of 0.67 µmol or less was effective in reducing the mutagenic effects of malon-aldehyde and β-propiolactone in certain tester strains of Salmonella typhimurium. Rosin & Stich (1979) showed that sodium selenite at concentrations of 3×10^{-4} and 1×10^{-3} mol caused a 50% inhibition of mutagenesis in S. typhimurium induced by N-methyl-N'-nitro-N-nitrosoguanidine or N-acetoxy-2-acetylaminofluorene, respectively. However, 10^{-2} mol sodium selenite was toxic to the bacteria in the presence of either carcinogen. Sodium selenite has also been shown to suppress spontaneous mutagenesis in yeast cultures (Rosin, 1981).

Jacobs et al. (1977b) examined the effects of sodium selenite on the mutagenicity of 2-acetylaminofluorene (AAF), N-hydroxy-2-acetyl-aminofluorene (N-OH-AAF), and N-hydroxy-aminofluorene (N-OH-AF) in the S. typhimurium TA 1538 bacterial tester strain. Metabolism of AAF and N-OH-AAF to the active mutagen, N-OH-AF was accomplished by rat liver extracts. Sodium selenite at concentrations ranging from 0.1 to 40 mmol decreased the mutagenicity of these compounds (Table 44). However, in the case of AAF, the authors noted that further decreases in mutagenicity induced by concentrations of selenium higher than 40 mmol were not tested, partly because of the formation of a red selenium compound of low solubility, which can be presumed to have been elemental selenium.

Ray et al. (1978) determined the frequencies of sister chromatid exchange (SCE) in lymphocytes resulting from the simultaneous exposure of whole-blood cultures to different concentrations of sodium selenite plus 10^{-4} mol methyl methanesulfonate (MMS) or 2.7×10^{-5} mol N-hydroxy-2-acetyl-aminofluorene (N-OH-AAF). The SCE frequency observed as a

Table 43. Effects of sodium selenite and various antioxidants on chromosomal breakage induced by 7,12-dimethylbenz(α)anthracene (DMBA) in human blood leukocyte cultures[a]

DMBA added (μmol)	Antioxidant added	Number of cells	Cells with breaks number	%	Breaks minus control (%)	Reduction in breaks (%)
0	none	211	23	10.9	-	-
1.6	none	290	82	28.3	17.4	-
1.6	0.20 μmol Na$_2$SeO$_3$	171	37	21.6	10.1	42.0
1.6	10 μmol dl-α tocopherol	156	28	17.9	6.4	63.2
1.6	10 μmol ascorbic acid	127	30	23.6	11.9	31.7
1.6	0.21 μmol butylated hydroxytoluene	157	29	18.4	6.3	63.8

[a] Adapted from: Shamberger et al. (1973a).

Table 44. Effect of sodium selenite on the mutagenicity of 2-acetylaminofluorene and its derviatives to Salmonella typhimurium[a]

Compound	Selenium added (mmol)	His$^+$ revertants per plate (± SD)	Mutagenic activity (% of appropriate control)
4.5 mmol AAF	-	1768 ± 149	100
	4	1411 ± 41	80
	10	1336 ± 69	76
	40	1148 ± 71	65
0.45 mmol N-OH-AAF	-	2353 ± 12	100
	0.4	1891 ± 210	80
	4	1598 ± 71	68
	10	1247 ± 53	53
	40	655 ± 43	28
0.065 mmol N-OH-AF	-	1628 ± 41	100
	0.1	1280 ± 88	79
	10	1233 ± 31	76
	20	999 ± 26	61

[a] Adapted from: Jacobs et al. (1977b).

result of coexposure to 2 compounds was less than that expected because of an additive SCE frequency response to each individual compound (Table 45). In their interpretive analysis of the data, the authors concluded that the most consistent explanation for the results was that sodium selenite decreased the SCE-inducing abilities of MMS and N-OH-AAF.

Martin et al. (1981) found that sodium selenite could protect against the mutagenic effects of acridine orange and 7,12-dimethylbenz[α]anthracene (DMBA) in the Ames Salmonella/microsome mutagenicity test. However, Chatterjee & Banerjee (1982) showed that the influence of sodium selenite on the transformation of mouse mammary cells, induced by DMBA added in organ culture, was markedly affected by the level of selenium used. For example, sodium selenite at concentrations of 10^{-7} - 10^{-8} mol increased the transformation frequency of the cells within the glands. At 10^{-5} mol, sodium selenite caused an 18 and 84% inhibition of the frequency of transformed cells at the initiation and promotional stages, respectively. At 10^{-4} mol, sodium selenite was toxic to the cells in vitro. Sodium selenite inhibited the metabolism and mutagenicity of benzo(a)pyrene to S. typhimurium strain TA 100 in rat (Teel & Kain, 1984) and hamster (Teel, 1984) liver S9 activation systems, but had no inhibitory effect on the mutagenicity of 1,2-dimethylhydrazine or azoxymethane for S. typhimurium G46 in the host-mediated assay (Moriya et al., 1979). On the other hand, sodium selenite at 3.95 x 10^{-9} - 1.98 x 10^{-8} mol protected against chromosomal damage in cultured human lymphocytes caused by 2.3 x 10^{-6} mol sodium arsenite (Sweins, 1983). However, this author reported that the cytotoxic concentration of sodium selenite in his system was very low (somewhere between 1.98 and 3.95 x 10^{-8} mol). No explanation was offered for the considerable variations between laboratories with respect to selenium cytotoxicity.

Several investigators have now examined the anti-mutagenic effects of dietary selenium in a variety of experimental systems. For example, Gairola & Chow (1982) fed rats either a low-selenium diet based on Torula yeast or the same diet supplemented with sodium selenite at 2 mg selenium/kg for 20 weeks. Dietary selenium did not have any effect on the metabolic activation potential of S9 liver enzymes towards benzo(a)pyrene in the Ames Salmonella/microsome mutagenicity assay. S9 mixtures from selenium-deficient rats were more active towards 2-aminoanthracene and less active towards 2-aminofluorene than mixtures from the selenium-supplemented rats. The authors concluded that further studies were needed to elucidate the role and nature of in vitro metabolites causing mutations in the bacteria. In contrast, Schillaci et al. (1982) did not find any differences in the mutagenic

Table 45. Effect of methyl methanesulfonate (MMS) or N-hydroxy-2-acetylaminofluorene (N-OH-AAF), with or without different concentrations of sodium selenite, on sister chromatid exchange (SCE) frequencies in whole-blood cultures of lymphocytes[a]

Compound	Na2SeO3 (mol)	Total cells scored	Observed SCE/cell	Expected SCE/cell[b]	Observed/ expected (%)
None	none	100	6.74 ± 0.30	–	–
10⁻⁴ mol MMS	none	100	30.17 ± 0.75	–	–
	1.58 x 10⁻⁶	75	30.60 ± 0.74	30.60 (0.43)	100
	3.95 x 10⁻⁶	75	30.13 ± 0.96	30.92 (0.75)	97
	7.90 x 10⁻⁶	75	33.15 ± 1.15	44.14 (13.97)	75
	1.19 x 10⁻⁵	75	31.60 ± 1.17	44.94 (14.77)	70
None	none	125	7.65 ± 0.27	–	–
2.7 x 10⁻⁵ mol N-OH-AAF	none	125	13.61 ± 0.43	–	–
	1.58 x 10⁻⁶	100	13.79 ± 0.42	13.61 (0.00)	100
	7.90 x 10⁻⁶	65	22.66 ± 1.10	27.15 (13.54)	83
	1.19 x 10⁻⁵	25	26.16 ± 1.86	29.24 (15.63)	89

[a] Adapted from: Ray et al. (1978).

[b] The expected SCE/cell was determined by adding the observed separate contributions to SCE frequency due to either MMS or N-OH-AAF to that due to different concentrations of Na2SeO3, as determined in a previous study (shown in parentheses (see also Table 42).

activation of 7-12-dimethylbenz[α]anthracene (DMBA) by liver
S9 mixtures in the Ames test with S. typhimurium strain TA 98,
prepared from rats fed Torula yeast-based diets supplemented
with sodium selenite at 0.1, 2.5, or 5.0 mg selenium/kg, from
weaning for 3 weeks. However, if the rats were injected with
20, 50, or 100 mg Aroclor® 1254/kg body weight, 5 days
before sacrifice, dietary selenium at 2.5 or 5.0 mg/kg in the
form of sodium selenite markedly decreased the mutagenic
activation of DMBA by liver microsomes.

Lawson & Birt (1983) measured single-strand breaks (SSB)
produced in pancreatic DNA by injecting 20 mg N-nitrosobis(2-
oxopropyl) amine (BOP)/kg body weight subcutaneously into
hamsters that had been fed a Torula yeast-based diet supplemented with sodium selenite at 0, 0.1, or 5 mg selenium/kg,
for 4 weeks previously. One hour after injection with BOP,
there were 2.26 ± 0.47, 2.83 ± 0.43, and 1.74 ± 0.43 SSB
per 10^8 daltons of DNA (mean ± SEM), respectively, in the
3 dietary groups, and the approximate half-lives of the SSB
were 33, 30, and 8 days, respectively. On this basis, the
authors suggested that high levels of dietary selenium may
stimulate the repair of carcinogen-induced DNA damage.

Olsson et al. (1984) used a novel isolated rat liver/cell
culture system to study the effects of selenium deficiency and
selenium supplementation, within the physiological range, on
the anti-mutagenic effects of the element. Rats were first
fed a Torula yeast-based selenium-deficient diet for 5 - 6
weeks with or without sodium selenite supplementation at
0.2 mg selenium/litre drinking-water. The rat livers were
then connected to an isolated liver perfusion system. A glass
plate with cultured Chinese hamster V79 cells was introduced
into the perfusion system immediately before the addition of
5 mmol dimethylnitrosamine (DMN) and then exposed directly to
the circulating perfusate. After 2 h, this glass plate was
exchanged for another with fresh V79 cells, which were exposed
during the subsequent 2 h. This procedure was adopted to
avoid a possible toxic effect that might influence mutation
frequency. As in vitro controls, V79 cells were treated for
2 h in Krebs-Ringer albumin solution with or without DMN. The
induced mutation frequencies in the 2 successively treated
cell populations were summarized to give the value for each
liver. It was found that the mutagenicity of DMN in Chinese
hamster V79 cells, after metabolic activation by the isolated
perfused rat liver, was approximately doubled when livers from
selenium-deficient rats were used compared with livers from
rats given the supplementary selenium. The authors were
unable to provide a precise biochemical explanation for the
mechanism by which selenium deficiency increased the mutagenicity of DMN, but they noted that microsomal N-oxygenation of
N,N-dimethylaniline (DMA) was decreased in livers from

selenium-deficient rats and suggested that further investigation of the different enzymes involved in DMA-$\underline{N}$-oxygenation appeared warranted. Another application of this liver perfusion/cell culture system was demonstrated by the work of Beije et al. (1984) who showed that bile collected from selenium-deficient livers perfused with a medium containing 5 mmol 1,1-dimethylhydrazine was much more mutagenic toward Chinese hamster V79 cells than bile from livers of rats given supplementary selenium and perfused in a similar manner. The authors suggested that the increased biliary excretion of reactive mutagenic metabolites observed in their selenium-deficient rat liver perfusion system might furnish a potential explanation for some of the protective effects of selenium reported against chemically-induced colon cancer in experimental animals.

7.6 Teratogenicity

Franke & Tully (1935) obtained chicken eggs from 2 farms in South Dakota that had histories of low hatchability. Hatchability in the 2 test groups was extremely low, being 4% in one and 12% in another. Examination of the eggs revealed that about 75% of those that had failed to hatch on the 21st day contained deformed embryos. Franke et al. (1936) then showed that injection of selenium salts into the air cell of eggs before incubation resulted in monsters similar to those occurring naturally on affected farms. Selenium injected as selenite to give a final concentration in the egg of 0.6 mg selenium/kg was the most effective in this respect, higher doses causing embryonic death and lower doses producing fewer monsters (Table 46). When the toxicity of various selenium compounds for chick embryos was compared, it was found that selenate was about twice as toxic as selenite (Palmer et al., 1973) (Table 47). The toxicity of selenomethionine was about the same as that of selenate but methylseleninic acid was more than twice as toxic. The toxicity of trimethylselenonium chloride was relatively low. The most common embryonic deformities observed in this study were the underdevelopment of the beak and abnormal development of the feet and legs, especially a fusing or webbing of the 2 outside toes. Because of the metabolic relationships between selenium and arsenic or cadmium, Holmberg & Ferm (1969) investigated the teratogenic potential of these elements, administered separately or together to golden hamsters (Table 48). Embryonic malformations were not observed when pregnant hamsters were injected intravenously with a barely sub-lethal dose of sodium selenite alone, and the 6% resorption rate observed was stated to be similar to the normal rate of resorption in this species. Furthermore, under these conditions, sodium selenite actually

Table 46. Teratogenicity of sodium selenite injected into hens' eggs[a]

Sodium selenite (mg selenium/kg)	Total embryos	Dead (%)	Abnormal (%)	Normal (%)
0.9	5	60.0	20.0	20.0
0.8	16	12.5	31.0	56.5
0.7	4	50.0	25.0	25.0
0.6	4	50.0	50.0	0
0.5	12	50.0	8.3	41.7
0.1	131	24.4	2.3	73.3
0.02	78	53.8	3.8	42.3
0.01	64	34.3	9.4	56.3

[a] Adapted from: Franke et al. (1936).

gave partial protection against the teratogenesis induced by sodium arsenate or cadmium sulfate.

7.7 Carcinogenicity and Anti-Carcinogenicity

7.7.1 Selenium as a possible carcinogen

Five investigations have been reported in the literature on the alleged carcinogenicity of selenium for laboratory animals. Nelson et al. (1943) fed groups of 18 female rats of an inbred Obsorne-Mendel strain a 12% protein diet consisting of 49% corn, 44% wheat, 3% yeast, and 1% each of cod liver oil, calcium carbonate, sodium chloride and dried whole liver. The diet was supplemented with 0, 5, 7, or 10 mg selenium/kg, as seleniferous corn or wheat, or 10 mg selenium/kg, as a mixed inorganic selenide containing ammonium potassium selenide and ammonium potassium sulfide. Although cirrhosis was frequently observed after 3 months of selenium exposure, no tumours or advanced adenomatoid hyperplasia were seen in any of the 73 selenium-exposed rats that died or were sacrificed before 18 months. Of the 53 selenium-exposed rats that survived 18 - 24 months, 43 had cirrhosis and 11 developed liver cell adenoma or low-grade carcinoma without metastasis in cirrhotic livers. There was no relationship between the extent of cirrhosis and tumour incidence except

Table 47. Comparative toxicity of various selenium compounds
for chick embryos[a]

Compound	Dose (mg selenium/kg)	Livability ratio (live chicks/ fertile eggs)	LD_{50} with 95% confidence limits (mg selenium/kg)
sodium selenite	0.0	16/18	
	0.05	18/18	
	0.1	16/17	0.3[b]
	0.2	17/18	
	0.4	2/18	
	0.8	0/18	
sodium selenate	0.0	18/18	
	0.1	12/18	
	0.2	4/18	0.13
	0.4	1/18	(0.086 -
	0.8	0/18	0.17)
	1.6	0/18	
selenomethionine	0.0	17/18	
	0.05	18/18	
	0.1	13/18	0.13
	0.15	5/18	(0.095-0.15)
	0.2	4/18	
	0.4	0/18	
methylseleninic acid	0.0	23/24	
	0.025	17/18	
	0.05	9/18	0.052
	0.1	2/24	(0.039-0.065)
	0.2	0/24	
	0.4	0/24	
trimethylselenonium chloride	0.0	18/18	
	6.0	17/18	
	12.0	13/18	15.7
	18.0	5/17	(12.7 -
	24.0	5/18	19.0)
	30.0	3/18	

[a] Adapted from: Palmer et al. (1973).
[b] An estimate since the data did not allow calculation by the computer programme.

that there were no tumours in any of the 18- to 24-month-old rats that did not have cirrhosis. The incidence of spontaneous adenoma and low-grade carcinoma of the liver in the unexposed rats was low and the incidence of spontaneous hepatic tumours in the rat colony was less than 1% in 18- to 24-month-old rats.

Table 48. Effect of selenium, arsenic, and cadmium on embryonic death and malformations in the golden hamster[a]

Treatment	Dose (mg/kg)	Total number of embryos	Number of embryos resorbed	Number of embryos malformed	Malformed (%)	Malformed or resorbed (%)
sodium selenite	2	86	5	0	0	6
sodium arsenate	20	177	62	86	49	84
sodium arsenate + sodium selenite	20 2	144	28	28	19	39
cadmium sulfate	2	115	24	59	51	72
cadmium sulfate + sodium selenite	2 2	82	2	3	4	6

[a] Adapted from: Holmberg & Ferm (1969).

The results of some of the initial studies of Volgarev &
Tscherkes (1967) seemed to indicate an increased incidence of
tumours in rats fed a low-protein diet supplemented with
sodium selenate at 4.3 mg selenium/kg, but subsequent tests
carried out by these authors failed to confirm these results.
Moreover, their work suffered from the fact that no control
groups, consuming diets without additional selenium, were
included in any of their trials. Schroeder & Mitchener
(1971b) reported an increased incidence of tumours in rats
given sodium selenate at 2 mg selenium/litre in the drinking-
water for the first year of life followed by 3 mg/litre until
death. In a previous evaluation of this work (US NAS/NRC,
1976), it was observed that the selenate-treated rats survived
longer than the untreated rats and this could have contributed
to the increased tumour incidence observed in the rats
receiving selenate. Also, as the organs and tissues did not
appear to have been systematically searched, the type and
incidence of histological lesions could not be known with
certainty.

Schroeder & Mitchener (1972) carried out 2 studies in
which mice were given selenium in the drinking-water. In both
studies, the mice were fed a diet composed of 60% whole rye
flour, 30% dried skim milk, 9% corn oil, and 1% iodized sodium
chloride, to which were added vitamins and iron. The diet was
calculated to contain 24% protein, 65% carbohydrate, and 11%
fat (dry weight). The mice were given doubly deionized water
for drinking, which originally came from a forest spring.
Certain essential trace metals were added to the drinking-
water, as soluble salts, in the following concentrations
(mg/litre): zinc, 50; manganese, 10; copper, 5; cobalt, 1;
and molybdenum 1. All mice received these trace elements in
their drinking-water. In addition, chromium was added to the
drinking-water at a level of 1 or 5 mg/litre in the first and
second studies, respectively. In the first study, groups of
100 or more Swiss mice of the Charles River CD strain,
containing equal numbers of each sex, were given, at weaning,
sodium selenite at either 0 or 3 mg selenium/litre of the
trace element-fortified drinking-water. This regimen
continued over the entire life span of the mice. The second
study was identical to the first, apart from the level of
chromium given in the drinking-water, and the fact that the
selenium was administered in the form of sodium selenate. Of
the 180 control mice autopsied from both studies, 119 were
sectioned. Tumours were found in 23 (19%) of those sectioned
and 10 of the tumours (43%) were malignant. The different
forms of selenium given did not influence the incidence of
tumours and, of the 176 selenium-exposed mice autopsied in
both studies, 88 were sectioned. Tumours were found in 13
(15%) of those sectioned but all tumours were malignant. It

was concluded that selenium had little tumourigenic or carcinogenic activity in mice, though, when tumours did appear in the selenium-exposed mice, they were all malignant.

The results of studies of Harr et al. (1967) (section 7.1.2.2) failed to demonstrate any tumours attributable to selenium, but most of the rats fed semi-purified diets containing levels of selenium greater than 2 mg/kg died within 100 days and almost all were dead before 2 years. Exceptions included 1 rat fed selenate at 4 mg selenium/kg diet containing 12% casein and 0.3% DL-methionine, and 27 rats alternating at weekly intervals between a control ration and a diet containing sodium selenate at either 4 or 8 mg selenium/kg. However, no hepatic tumours were observed, even in the 71 rats that survived 2 years or longer at continuous dietary-selenium levels of 0.5 - 2.0 mg/kg.

The above studies indicate that test animals develop neoplastic lesions, only when they have liver cirrhosis produced by frank selenium toxicity (i.e., no hepatomas were observed in the absence of severe hepatotoxic phenomena). For this reason, it was concluded that selenium was not, by reason of its capacity to induce liver damage when consumed at high levels, properly classified as carcinogenic because of its potential association with a higher rate of liver cancer (Gardner, 1973).

Jacobs & Forst (1981b) did not observe any signs of neoplasia in groups of 35 female Swiss mice fed a commercial pelleted diet and given 0, 1, 4, or 8 mg selenium/litre drinking-water, as sodium selenite, for 50 weeks.

Three studies have shown carcinogenic effects that appear to be more a specific effect of a particular selenium compound rather than an effect of selenium itself. Seifter et al. (1946) found that 8 white rats that had received 0.05% bis-4-acetamino-phenyl-selenium dihydroxide in their diet for 105 days had multiple adenomas of the the thyroid glands and adenomatous hyperplasia of the liver. Innes et al. (1969) fed the maximal tolerated dose of selenium diethyldithiocarbamate (Ethyl selenac) to a group of 72 mice containing both sexes of two hybrid strains (C57BL/6 x C3H/Anf)F and (C57BL/g x AKR)F_1. The mice were given this substance by stomach tube at a dose of 10 mg/kg body weight, starting at 7 days of age. At 4 weeks of age, the chemical was mixed directly into the diet at a concentration of 26 mg/kg. After 82 weeks, all the surviving mice were sacrificed; of 69 necropsied, 26, 13, and 5.8% had hepatomas, lymphomas, and pulmonary tumours, respectively. Among 338 negative control mice, most of which were sacrificed between 78 and 89 weeks, the incidences of the corresponding tumours were 4.1, 4.1, and 6.2%, respectively.

A report from the US National Cancer Institute (National Cancer Institute, 1980) suggested that commercial selenium sulfide, an ingredient in certain anti-dandruff shampoos, was carcinogenic for rats and mice. Elemental analysis of the test chemical used in this study indicated that the material was a mixture of selenium monosulfide and selenium disulfide. The melting point of the test sample was closer to that reported for the monosulfide than that reported for the disulfide and the X-ray diffraction pattern was consistent with patterns reported for the monosulfide. It was concluded that the selenium in the test substance used in this bioassay was present primarily as the monosulfide.

In the rat study, groups of 4-week-old male and female Fischer F344 rats were fed presterilized lab meal and were given untreated well water ad libitum for 104 - 105 weeks. During the first 103 weeks, 50 rats of each sex were given one of 4 treatments: untreated control; vehicle control (received volumes of 0.5% aqueous carboxymethylcellulose equal to those of the test solutions administered); low-dose group (3 mg selenium sulfide suspended in 0.5% aqueous carboxymethyl-cellulose/kg body weight, 5 days per week, given by gavage); or high-dose group (15 mg selenium sulfide/kg body weight administered by the same route and schedule). The results showed an increased incidence of primary liver tumours in both male and female high-dose groups (Table 49).

Neoplastic nodules were usually single, rather well-defined areas characterized by altered hepatocytes. In most instances, the hepatocytes were larger than normal, eosinophilic, and occasionally vacuolated. The normal architecture was altered, mainly resulting in a solid mass of hepatocytes or a trabecular pattern rather than the normal hepatic cords. The mass compressed the adjacent parenchyma around the periphery. Anaplasia and mitoses were minimal. Hepatocellular carcinomas were usually large multinodular masses, often encompassing entire liver lobes or even multiple lobes. The histological appearance of these neoplasms varied from areas appearing normal to much more anaplastic areas. The neoplastic hepatocytes varied from small basophilic cells to very large eosinophilic and occasionally vacuolated cells. Mitoses were variable. No distant metastases were observed in any of the rats bearing hepatocellular carcinomas. No other neoplasms were found that could be related to the administration of the selenium sulfide.

An increased incidence of focal cellular change in the liver was noted in high-dose male rats but was essentially comparable in frequency in the remaining treated and control groups of each sex.

A compound-related increase in pigmentation in the lungs was observed. This was characterized by the accumulation of dark, slightly granular-appearing pigment in the interstitial areas and in some peribronchial areas. In most cases, the

Table 49. Effect of selenium sulfide, given orally for 103 weeks, on the incidence of liver tumours in Fischer F344 rats[a]

Test group	Sex	Initial number of rats	Selenium sulfide dose	Tumour incidence	
				Hepatocellular carcinoma	Neoplastic nodules
			(mg/kg)		
Untreated control	M	50	0	1/48 (2%)	3/48 (6%)
Vehicle control[b]	M	50	0	0/50 (0%)	1/50 (2%)
Low-dose[c]	M	50	3	0/50 (0%)	0/50 (0%)
High-dose[c]	M	50	15	14/49 (29%)	15/49 (31%)
Untreated control	F	50	0	0/50 (0%)	0/50 (0%)
Vehicle control[b]	F	50	0	0/50 (0%)	1/50 (2%)
Low-dose[c]	F	50	3	0/50 (0%)	0/50 (0%)
High-dose[c]	F	50	15	21/50(42%)	25/50 (50%)

[a] Adapted from: National Cancer Institute (1980).
[b] Received only vehicle for dosing (0.5% aqueous carboxymethylcellulose) 5 days per week, by gavage.
[c] Received stated dose of selenium sulfide suspended in 0.5% aqueous carboxymethylcellulose, 5 days per week, by gavage.

pigment appeared to be located within cells, principally macrophages. No evidence of inflammation, relative to the pigment deposits, was noted. Lung pigmentation was found in 47/49 (96%) high-dose males, 1/50 (2%) low-dose males, 45/50 (90%) high-dose females, and 36/50 (72%) low-dose females, but not in control males or females.

The protocol for the testing of selenium sulfide in B6C3F1 mice was similar to that used for rats, except that the doses of the substance under test were increased (Table 50). There was an increased incidence of primary liver and lung tumours in the high-dose female mice and a marginal increase in the incidence of these tumours in high-dose males.

Hepatocellular carcinomas varied from single nodules to multinodular masses, often encompassing several liver lobes. Individual hepatocytes varied considerably in morphology from large eosinophilic cells to small, darkly-staining hepatocytes. In many cases, there was marked variation in cell type in different parts of the neoplasm. The number of mitoses varied. The number of hepatocellular carcinomas metastasizing to the lungs was comparable in vehicle-control and high-dose males. No metastases were observed in the lung in female mice. Alveolar/bronchiolar adenomas were usually small solitary lesions located in the subpleural area or immediately adjacent to a bronchiole. The cells involved were cuboidal to tall columnar and tended to be situated perpendicular to the basement membrane in a single layer. These cells were arranged in complex papillary projections forming discrete nodules and compressing adjacent alveolar walls. Mitoses were rare. Alveolar/bronchiolar carcinomas, however, were less discrete lesions and tended to be larger and occasionally multiple, consisting of a confluence of two or more nodules. The individual cells tended to be less rigidly arranged along basement membranes and were often piled up in layers or arranged in solid sheets without a papillary pattern. The cells often showed increased basophilia and a moderate mitotic index. Evidence of invasion into adjacent vessels, or extension into bronchioles and adjacent lung parenchyma was frequently present.

Other neoplasms that occurred in the mice were similar in number and kind to those that usually occur in aged B6C3F1 control mice and could not be related to the long-term administration of selenium sulfide.

In a comparison study, the effect of the dermal application of selenium sulfide was examined. No increased incidence of neoplasms was observed that could be attributed to selenium sulfide.

Table 50. Effect of selenium sulfide given orally for 103 weeks on the incidence of liver and lung tumours in B6C3F1 mice[a]

Test group	Sex	Initial number of mice	Selenium sulfide dose (mg/kg)	Liver tumour incidence		Lung tumour incidence	
				hepatocellular carcinoma	hepatocellular adenoma	Alveolar/ bronchiolar carcinoma	Alveolar/ bronchiolar adenoma
Untreated control	M	50	0	17/49 (35%)	3/49 (6%)	1/49 (2%)	8/49 (16%)
Vehicle control[b]	M	50	0	15/50 (30%)	0/50 (0%)	1/50 (2%)	3/50 (6%)
Low-dose[c]	M	50	20	11/50 (22%)	3/50 (6%)	2/50 (4%)	8/50 (16%)
High-dose[c]	M	50	100	23/50 (46%)	0/50 (0%)	2/50 (4%)	12/50 (24%)
Untreated control	F	50	0	2/50 (4%)	1/50 (2%)	0/50 (0%)	2/50 (4%)
Vehicle control[b]	F	50	0	0/49 (0%)	0/49 (0%)	0/49 (0%)	0/49 (0%)
Low-dose[c]	F	50	20	1/50 (2%)	1/50 (2%)	1/50 (2%)	2/50 (4%)
High-dose[c]	F	50	100	22/49 (45%)	6/49 (12%)	4/49 (8%)	8/49 (16%)

[a] Adapted from: National Cancer Institute (1980).
[b] Received only vehicle for dosing (0.5% aqueous carboxymethylcellulose), 5 days per week, by gavage.
[c] Received stated dose of selenium sulfide suspended in 0.5% aqueous carboxymethylcellulose, 5 days per week, by gavage.

7.7.2   Selenium as a possible anti-carcinogen

High levels of dietary selenium have been shown to protect laboratory animals against chemical carcinogenesis, under a wide variety of conditions. Clayton & Baumann (1949) fed 2 groups of 15 young adult rats, weighing approximately 200 g, a basal diet consisting of extracted casein, 12 parts; salts, 4 parts; corn oil, 5 parts; and glucose monohydrate (Cerelose) to 100 parts. The diet was supplemented with thiamin, riboflavin, pyridoxine, calcium pantothenate, and choline. Fat-soluble vitamins were provided by giving 2 drops of halibut liver oil per rat every 4 weeks. During the initial 4-week feeding period, both groups received the basal diet supplemented with 0.064% 3'-methyl-4-dimethylaminoazobenzene. This was followed by a 4-week "interruption period" during which neither group received the azo dye, but one group received the basal diet supplemented with sodium selenite at 5 mg selenium/kg, and the other group received the basal diet without any added selenium. During the third, 4-week feeding period, both groups received the basal diet supplemented with 0.048% of the azo dye, but no supplementary selenium. During the final 8-week feeding period, both groups received the basal diet without either azo dye or selenium. Two of the 9 surviving rats in the group that received supplementary selenium during the "interruption period" had liver tumours (22%), compared with 4 out of 10 survivors (40%) in the group not supplemented with selenium. The results of a second similar study showed a liver tumour incidence of 4/13 (31%) in the selenium-supplemented group compared with 8 out of 13 rats (62%) in the unsupplemented group.

A preliminary study by Shamberger & Rudolph (1966) indicated that concomitant dermal application of sodium selenide reduced the tumour-promoting effect of croton oil in mice initiated with 7,12-dimethylbenz[a]anthracene (DMBA). Riley (1968) found that a similar application of sodium selenide prevented the mast cell reaction caused by an active fraction of croton oil in the skin of DMBA-initiated mice. In a later set of studies (Shamberger, 1970), 2 groups of 30 female ICR Swiss mice, 50 - 55 days old, were treated once on day one with 0.125 mg DMBA dissolved in 0.25 ml acetone. On days 2 - 21, the shaved backs of the first group of mice were treated with 0.25 ml of a 20:80 water-acetone mixture containing 0.0005% sodium selenide, whereas the second group was not treated with any anti-oxidant. Subsequently, in 3 separate, but essentially identical, studies, both groups of mice received 0.25 ml of 0.04% croton oil in acetone daily for 16, 18, and 18 weeks respectively. After the croton oil treatment, the incidence of papillomas in the selenide-treated groups was 17, 63, and 45%, respectively, in 3 separate

studies, compared with an incidence in the non-treated groups
of 43, 63, and 89%, respectively. The corresponding number of
papillomas per mouse in the 3 studies was 1.5, 3.0, and 1.5,
respectively, in the selenide-treated groups and 3.2, 6.0, and
2.3 in the non-treated groups. No papillomas were observed in
3 DMBA-initiated control groups, which did not receive either
antioxidant or croton oil. A similar protective effect of
sodium selenide was observed in 3 additional studies in which
the promoting agents were croton oil, croton resin, and phenol
and in which the selenide was applied concomitantly with the
promoter.

In another study, 0.25 ml of 0.01% 3-methylcholanthrene
(MCA) was applied daily to the shaved backs of one group of
mice for 19 weeks; a second group received 0.0005% sodium
selenide together with the MCA. After 19 weeks, the incidence
of papillomas was 68% in the selenide-treated group and 87% in
the un-treated group, and the number of papillomas per mouse
was 3.2 and 2.2, respectively. After 30 weeks, the number of
mice with cancers was 17/28 and 25/30 in the selenide-treated
and untreated groups, respectively, the total number of
cancers per group being 29 and 71, respectively.

Shamberger (1970) also carried out 2 dietary studies in
which 4 groups of 36 female, albino ICR Swiss mice, 50 - 55
days old, were fed Torula yeast diets supplemented with sodium
selenite at 0, 0.1, or 1.0 mg selenium/kg or sodium selenide
at 0.1 mg selenium/kg. After 2 weeks on the test diets, 0.125
mg DMBA dissolved in acetone was applied once to the skin.
After 3 weeks, 0.25 ml 0.05% croton oil in acetone was applied
daily for 17 weeks. At this time, 26/36 mice fed the
unsupplemented diet had papillomas, compared with 14/35 mice
fed the diet supplemented with 1.0 mg selenium/kg, as
selenite. The incidence of papillomas was slightly elevated in
the mice fed the diet supplemented with 0.1 mg selenium/kg, as
sodium selenite, compared with the unsupplemented mice. The
incidence of papillomas in mice fed sodium selenide at
0.1 mg selenium/kg was intermediate between that of the unsupplemented group and the group fed sodium selenite at 1.0 mg
selenium/kg.

In a second study of similar design, 4 groups of 36 mice
were fed the test diets described above for 2 weeks. Then
0.25 ml of 0.03% benzo(a)pyrene in acetone was applied daily
to the shaved backs of the mice for 27 weeks. At this time,
the incidence of papillomas in the groups fed the torula diet
supplemented with 0, 0.1, or 1.0 mg selenium/kg, as selenite,
or 0.1 mg selenium/kg, as selenide, was 14/35, 22/36, 8/33,
and 12/35, respectively. The number of papillomas per mouse
in the corresponding groups was 8.1, 9.8, 5.0, and 6.8.

Harr et al. (1972) weaned 80 female OSU-Brown rats at 35
days of age and divided them into 4 groups of 20. Each group

was fed a low-selenium basal diet (0.018 µg selenium/kg)
that included Torula yeast as the protein source and contained
60 mg vitamin E/kg. Groups, 1, 2, 3, and 4 were fed the basal
diet supplemented with sodium selenite at 2.5, 0.5, 0.1, and 0
mg selenium/kg, respectively. All diets contained 150 mg
2-acetylaminofluorene/kg. The 40 rats in groups 1 and 2 were
born from parents reared on a selenium-depleted regimen, but
were clinically normal. The 40 rats in groups 3 and 4 were
from the second generation maintained on the depeletion
regimen and had clinical signs of selenium deficiency. After
200 days of treatment, the number of mammary adenocarcinomas
in groups 1, 2, 3, and 4 was 0, 1, 8, and 9, respectively.
After 320 days, the number of mammary adenocarcinomas in the
same groups was 11, 11, 13, and 12, respectively. Mammary
adenocarcinomas in groups 1, 2, and 3 occurred mainly (90%) in
the thoracic region and were well circumscribed, firm, and
easily removed, whereas those in group 4 occurred mainly (80%)
in the pelvic area and were soft and fluid, contained little
connective tissue, and were invasive. After 240 days of
treatment, the number of hepatic carcinomas in groups 1, 2, 3,
and 4 was 0, 1, 9, and 4, respectively. After 320 days, the
number of hepatic carcinomas in the same groups was 8, 12, 12,
and 6, respectively. Toxic hepatitis was observed in 18 of
the 20 livers from group 1 (selenium added at 2.5 mg/kg), but
was not seen in the other groups.

In studies of Marshall et al. (1978), 2 groups of male
albino Sprague Dawley rats were fed diets containing 2-acetyl-
aminofluorene at 0.3 g/kg, for 14 weeks. One group received 4
mg selenium, as sodium selenite/litre drinking-water. The
carcinogen was withdrawn from the diet for an additional 4
weeks and then the rats were sacrificed. The rats given
selenium had about 50% fewer liver tumours than unsupplemented
rats. Control rats given a similar level of selenium in the
water showed normal growth response, liver weight, and
appearance.

Three groups of 15 male Sprague Dawley rats, weighing
about 250 g, were fed a basal diet of finely ground Purina
Laboratory Chow and water ad libitum (Griffin & Jacobs, 1977).
One group did not receive any supplementary selenium, a second
group received 6 mg selenium, as sodium selenite/litre
drinking-water, and a third group received 6 mg selenium, in
the form of a high selenium yeast/kg diet. After one week on
this regimen, all 3 groups were given 0.05% 3'-methyl-4-
dimethyl-aminoazobenzene in the diet for 8 weeks. Following
this, all groups were maintained on the carcinogen-free diets
for an additional 4 weeks. The selenium supplements were
maintained for the second and third group throughout the
entire study. At sacrifice, the incidence of liver tumours in
the surviving rats was 92% (11/12) in the unsupplemented

group, 46% (7/15) in the group supplemented with selenite in the drinking-water, and 64% (9/14) in the group supplemented with selenium yeast in the diet. In this study, the selenium supplements had little effect on the growth of the rats. Histopathological examination of several of the tumours revealed that they were bile duct adenocarcinomas and liver cell adenocarcinomas.

Jacobs et al. (1977a) injected 2 groups of 15 male, 8-week-old Sprague Dawley rats, weekly, with 20 mg sym,-dimethylhydrazine dihydrochloride (DMH)/kg body weight, for 18 weeks. One group of rats received sodium selenite at 4 mg selenium/ litre drinking-water, which was available ad libitum one week prior to, and throughout, administration of the carcinogen. The other group did not receive selenium added to the drinking-water. The group receiving DMH and no added selenium had an 87% incidence (13/15) of colon tumours, whereas the group receiving both DMH and added selenium had a 40% incidence (6/15) of colon tumours. The total number of colon tumours was 39 in rats receiving only DMH and 11 in rats receiving both DMH and added selenium. In a similar study in which 2 groups of 15 rats were injected weekly with 20 mg of (methylazoxyl)-methanol acetate (MAM)/kg body weight for 18 weeks, no significant differences in the incidence of colon tumours were apparent between groups with (14/15) and without (14/14) added selenium in the drinking-water. However, the number of MAM induced tumours was 73 in the group given MAM alone and 42 in the group receiving both MAM and added selenium.

Schrauzer and colleagues carried out a series of studies in which exogenous carcinogens were not used, to test the ability of supplemental selenium in influencing the development of spontaneous mouse mammary tumours, presumably of viral origin. In study 1 (Schrauzer & Ishmael, 1974), 2 groups of 30 virgin female C$_3$H/St mice, 4 - 6 weeks old, were fed a basal diet ("Concord Maid") consisting of meat scraps, dried skimmed milk, oat groats, ground wheat, wheat germ meal, vegetable oil, cane molasses, salt, brewer's yeast, cereal binder, sodium propionate, and calcium pantothenate. The diet contained about 150 g protein, 5 g fat, and 0.15 mg selenium/kg. One group of mice received plain distilled water for drinking, while the other group received distilled water fortified with 2 mg selenium, as selenium dioxide/litre. The strain of mice used in this study ordinarily has a high incidence of spontaneous mammary adenocarcinomas and, after 16 months of treatment, the observed incidence in the group not given supplemental selenium in the water was 82%, whereas the incidence in the group given selenium was 10%.

In study 2 (Schrauzer et al., 1976), 3 groups of 30, 30, and 50 female C$_3$H/St mice fed the basal diet described above were also given selenite at 0, 5, or 15 mg selenium/litre drinking-water, respectively. Toxicity due to the doses of selenium was indicated by the average body weights in the 3 groups at 12 months (33, 29, and 25 g, respectively) and by a higher tumour-unrelated mortality rate in the selenium-exposed mice. After 26 months, the corresponding incidence of spontaneous mammary tumours in the 3 groups was 82, 36, and 33% in the mice that survived to the age at which the first tumours appeared in each group. A fourth group of 20 mice, fed a selenium-deficient diet for 14 months and then the basal diet until death, had a mammary tumour incidence of 69% which was not judged to be different from the 82% incidence in the control group (basal diet throughout life span and no added selenium in the water).

In study 3 (Schrauzer et al., 1978a), 4 groups of 30 female C_3H/St mice were fed a basal diet ("Wayne F-6 Lab Blox") that was different from that used in the first 2 studies and consisted of fish meal, animal liver meal, soybean meal, corn and wheat flakes, ground corn, wheat red dog, wheat middlings, soybean oil, cane molasses, salt, brewer's yeast, and various vitamins and minerals. This diet contained 244.8 g protein, 64.8 g fat, and 0.45 mg selenium/kg. The 4 groups received selenium dioxide at 0, 0.1, 0.5, or 1.0 mg selenium/litre drinking-water and, after 22 months, the incidence of mammary tumours was 42, 25, 19, and 10%, respectively. Selenium supplementation at these levels did not have any noticeable adverse effects on weight-gain or survival of the mice. The incidence of spontaneous mammary tumours in the control (unsupplemented) group was lower in this study than that observed in the 2 previous studies (42 compared with 82%), and this was attributed to the higher selenium content of the new basal diet used in the third study (0.45 versus 0.15 mg/kg).

Medina & Shepherd (1980) fed BALB/cfC3H mice Wayne Lab Blox and gave them selenium dioxide at 0, 2, or 6 mg selenium/litre drinking-water, ad libitum, starting at 10 weeks of age and continuing to the end of the study. In 12-month-old mice, 2 and 6 mg selenium/litre decreased mammary tumour incidence from 82% in the untreated controls to 48 and 12%, respectively. There were no effects of the selenium treatment on normal reproductive function or weight gain in these mice. In another study, samples of 4 different pre-neoplastic outgrowth lines (D2, C3, C4, and CD-7) were transplanted into the mammary gland-free fat pads of syngenic mice. When the implants had filled the mammary fat pads with their respective outgrowths (8 weeks after implantation), the mice were given selenium dioxide at 4 mg selenium/litre drinking-water, ad libitum, for the rest of the study. Such treatment with selenium delayed the rate of tumour formation only in line C4, increasing the time for half of the outgrowths to produce tumours from 34 to 44 weeks. In a third study, 2 - 6 mg selenium/litre drinking-water had no effect on the growth rate of primary tumours transplanted subcutaneously in BALB/c mice. Since selenium did not have any effect on tumour formation rate in 3 of 4 preneoplastic mammary outgrowth lines or on the growth rate of established mammary tumours, the authors suggested that selenium might act by inhibiting chemical or viral transformation of normal cells or by inhibiting expression of initially transformed cells.

Because of the metabolic antagonism between arsenic and selenium (section 7.1.6.3), Schrauzer and co-workers also investigated the effect of arsenic on the genesis of the spontaneous mammary tumours in C_3H/St female mice described above. In their first study on arsenic and selenium (which was part of the same study described above in Schrauzer & Ishmael (1974)), giving sodium arsenite at 10 mg arsenic/litre drinking-water to a group of 30 C_3H/St mice for 16 months, reduced the incidence of spontaneous mammary tumours to 27% compared with an incidence of 82% in the untreated controls (Schrauzer & Ishmael, 1974). However, arsenic treatment markedly stimulated the growth rate of spontaneous or transplanted mammary tumours. In a second study concerning arsenic, sodium arsenite at 80 mg arsenic/litre drinking-water, administered to a group of 20 C_3H/St mice, reduced the incidence of spontaneous mammary tumours to 40% compared with an incidence of 82% in the untreated controls (Schrauzer et al., 1976), but some of the tumour-inhibiting effect of arsenic appeared to be masked at this dose level by its toxicity.

In a third study relating arsenic and selenium (Schrauzer et al., 1978b), 4 groups of 30 female C_3H/St mice were fed the Wayne F-6 Lab Blox diet described above, which contained 0.29 mg arsenic/kg. The 4 groups received the following supplements in their deionized drinking-water: none, arsenic trioxide at 2 mg arsenic/litre, selenium dioxide at 2 mg selenium/litre, or 2 mg of both arsenic and selenium. The incidence of mammary tumours in the 4 groups was 41, 36, 17, and 62%, respectively, and the percentage of multiple mammary tumours was 17, 40, 0, and 28. The age of tumour onset in the corresponding groups was 4.5, 9, 16, and 8 months. Thus, in this case, treatment with arsenic appeared to diminish the cancer-protecting effect of selenium. Arsenic also accelerated tumour growth and increased the incidence of multiple tumours.

Newberne & Conner (1974) fed 4 groups of male rats of the Charles River CD strain, weighing about 100 g each, a basal diet consisting of casein, 200 g/kg; sucrose, 209 g/kg; dextrose, 209 g/kg; dextrin, 209 g/kg; stripped lard, 80 g/kg; Wesson Oil, 20 g/kg; selenium-free salts, 50 g/kg; vitamin

mix, 20 g/kg; choline, 3 g/kg; and vitamin B_{12}, 0.05 g/kg.
The basal diet contained about 0.03 mg selenium/kg. One group
received the unsupplemented basal diet, whereas the other 3
groups received the basal diet supplemented with sufficient
sodium selenite to attain approximate dietary levels of 0.1,
1.0, or 5.0 mg selenium/kg, respectively. Each group was fed
the diet for 2 - 3 weeks, before oral administration of 7 mg
aflatoxin B_1 in dimethyl sulfoxide/kg body weight. The rats
continued on their respective diets for an additional 2 weeks,
and the mortality rate after this time, in the groups fed the
diets containing 0.03, 0.1, 1.0, and 5.0 mg selenium/kg was
28/29, 20/30, 7/28, and 27/29, respectively. Histological
examination revealed that rats receiving 1.0 mg selenium/kg
diet were partially protected against the aflatoxin B_1 and
also had less severe liver lesions. However, the groups fed
diets containing 1.0 or 5.0 mg selenium/kg exhibited a novel
renal lesion associated with acute aflatoxin B_1 toxicity,
characterized by marked tubular necrosis at the cortical
medullary junction. Some tubules exhibited hyperplastic
changes of the epithelium as well as necrosis.

In a second study, Grant et al. (1977) gave 25 µg
aflatoxin B_1 orally, 5 days/week for 4 weeks, to 140 male
Sprague Dawley rats. The rats were fed normal or marginally
lipotrope-deficient semisynthetic basal diets containing
sodium selenite at 0.03, 0.10, 0.50, 1.0, 2.5, or 5.0 mg
selenium/kg. After 17 months, the surviving rats were
sacrificed and histopathological examination was carried out.
A very low incidence (20%) of hepatocarcinomas was seen in
rats receiving 1.0 mg selenium/kg in the normal basal diet and
0.10 or 0.50 mg/kg in the lipotrope-deficient diet. No other
tumours were observed, and grading of the hepatic lesions
indicated that there was no significant differences among the
dietary selenium groups. Dietary sodium selenite and repeated
doses of aflatoxin B_1 interacted to produce large bizarre
cells in the renal tubules, occurring in the same region of
the kidney as the severe necrosis seen previously in rats fed
high levels of dietary selenium and given an acute dose of
aflatoxin B_1.

Recently, a study of the effects of selenium on aflatoxin
B_1-induced enzyme altered foci in rat liver has been
reported (Milks et al., 1985). Male Sprague Dawley rats were
fed a selenium-deficient diet and given sodium selenite at 5,
2, 0.2, or 1 mg selenium/litre drinking-water for 3 weeks.
Each rat then received 2 µg mol aflatoxin B_1/kg body
weight by stomach tube. For the next week the selenium status
was "normalized". Rats previously receiving 5, 2, 0.2, or 0
mg selenium/litre received, respectively, 0. 0.2, 2, or 5
mg/litre. Then rat chow was fed and a promoting regimen
consisting of phenobarbital in the drinking-water, and a

partial hepatectomy was instituted. Eight weeks later, necropsies were carried out and livers were stained histochemically for γ-glutamyl transpeptidase activity. Foci of activity were counted. The number of foci seen in livers from rats given 5, 2, 0.2, or 0 mg selenium/litre during initiation were 0.62 ± 0.22, 1.97 ± 0.46, 3.35 ± 0.66, and 2.46 ± 0.23 per cm^2, respectively. The data suggested that 5 mg selenium/litre can protect against the hepatocarcinogenic effects of aflatoxin B_1 in the rat.

On the basis of anecdotal observations, Wedderburn (1972) suggested that a decrease in the number of cases of intestinal carcinoma in autopsied sheep might be associated with the widespread veterinary use of selenium to prevent deficiency diseases in sheep. Simpson (1972a) conducted an investigation into the epidemiology of carcinomas of the small intestine of sheep in which the viscera of 32 733 ewes were examined. Carcinomas of the small intestine were diagnosed in 483 animals, but the prevalence of these neoplasms in ewes regularly dosed with selenium throughout their lives was not significantly different from the prevalence in ewes never treated with selenium (Simpson, 1972b). Furthermore, no significant differences were found in association with differences in soil types on the farms from which the sheep originated. It was concluded that soil-selenium levels and administration of selenium in the form and at the dose rates used did not have any effects on the development of intestinal carcinomas in sheep. Underwood (1977) commented that neoplasias were not observed among the various lesions attributed to selenium deficiency in animals.

Ip has made and summarized a series of observations on selenium and carcinogenesis (Ip, 1985a).

Evidence of an interaction between dietary fat and selenium status in the induction of mammary tumours by dimethylbenz[α]anthracene in rats has been presented by Ip & Sinha (1981). They fed 8 groups of 23 - 25 female weanling Sprague Dawley rats one of the following diets either deficient in selenium or supplemented with sodium selenite at 0.1 mg selenium/kg: 1% corn oil, 5% corn oil, 25% corn oil, or 1% corn oil plus 24% hydrogenated coconut oil. In addition to the fat, the diets consisted of Torula yeast, dextrose, HMW salt mix, vitamin mix, alphacel, and DL-methionine. The diets were adjusted so that the intake of all nutrients would be the same, except for dextrose and fat, assuming that the rats would consume an equal number of calories. The corn oil used was stripped of tocopherol and the unsupplemented diet was deficient in selenium (less than 0.02 mg/kg by fluorometry). Mammary tumours were induced by the intragastric administration of 5 mg dimethylbenz[α]anthracene at 50 days of age. Increasing the dietary polyunsaturated-fat level (corn oil)

increased the tumour incidence in rats fed the selenium-
supplemented diets (Table 51). However, the high-saturated-
fat diet was much less active in stimulating tumourigenesis.
Selenium deficiency increased the tumour yield only in rats
fed the high polyunsaturated-fat diet (25% corn oil).
Increased tumour incidence due to selenium deficiency was not
seen in the rats fed the low-fat diets containing polyunsaturated fat (1 or 5% corn oil) or the high-fat diet
containing primarily saturated fat (1% corn oil with 24%
coconut oil). Selenium deficiency, however, did increase the
incidence of mammary tumours in rats fed a low-fat diet
containing polyunsaturated fat (1% corn oil), when larger
doses of dimethylbenz[α]anthracene were used (10 or 15 mg -
data not shown).

Table 51. Incidence of palpable mammary tumours in
dimethylbenz(α)anthracene-treated rats fed different levels and
types of fats in the diet, with or without selenium supplementation[a]

Fats in diet		Selenium in diet (mg/kg)	Incidence of palpable tumours 19 weeks after dimethylbenz(α)-anthracene administration	
Corn oil (%)	Coconut oil (%)		Number of rats	(%)
1	0	0	4/23	17.4
1	0	0.1	3/24	12.5
5	0	0	11/25	44.0
5	0	0.1	8/24	33.3
25	0	0	24/25	96.0
25	0	0.1	15/25	60.0
1	24	0	7/24	29.2
1	24	0.1	6/25	24.0

[a] Adapted from: Ip & Sinha (1981).

In a recent study, the effects of vitamin E status on the
anticarcinogenic effect of selenium were examined (Ip, 1985b).
Female Sprague Dawley rats were fed a 20% stripped corn oil,
casein-based diet. Four dietary groups were formed: adequate
vitamin E/adequate selenium, adequate vitamin E/high selenium,
deficient vitamin E/adequate selenium, and deficient vitamin
E/high selenium. Adequate and deficient vitamin E diets

contained 50 and 10 mg vitamin E/kg diet, respectively.
Adequate and high selenium diets contained Na_2SeO_3 at 0.1
and 2.5 mg selenium/kg, respectively. At 50 days of age, rats
received 5 mg of dimethylbenz[α]anthracene (DMBA) each by
stomach tube. Rats were killed 20 weeks after DMBA administration. The tumour incidences in the groups were: adequate
vitamin E/adequate selenium, 76%; adequate vitamin E/high
selenium, 40%; deficient vitamin E/adequate selenium, 84%;
deficient vitamin E/high selenium, 68%. This suggests that
selenium protection against carcinogenesis is decreased in
vitamin E deficiency.

Pence & Buddingh (1985) studied the effects of selenium
deficiency on 1,2-dimethylhydrazine (DMH)-induced colon cancer
in the rat. They used male Sprague Dawley rats fed Torula
yeast-based diets containing 2% corn oil. Sodium selenite at
0.1 mg selenium/kg was added to the diet of the controls.
Weanlings were fed the diets 3 weeks prior the institution of
DMH treatment and were killed after 20 weeks of treatment. No
effects of selenium status were noted on the incidence of
colon adenocarcinomas.

The effects of selenium on UVR-induced skin carcinogenesis
were studied by Overvad et al. (1985). Female hairless mice
were given sodium selenite at 0, 2, 4, or 8 mg selenium/litre
drinking-water. Three weeks after selenium exposure began
they were exposed to UVR daily for 22 weeks. Then they were
examined weekly for 26 weeks for skin tumours, and relative
tumour onset ratios were calculated. The 2 mg selenium/litre
treatment did not affect tumour onset but the higher doses
did. This suggests that selenium can protect against UVR-
induced skin cancer.

Birt et al. (1984) reported that high levels of selenium
in the diet increased pancreatic carcinogenesis induced by
bis-(2-oxopropyl)-nitrosamine (BOP) in male Syrian hamsters.
Torula yeast-based diets with either 0.1 or 2.5 mg selenium/kg
were fed to hamsters beginning at 4 weeks of age. BOP was
given in 4 weekly injections of 5 mg/kg body weight, beginning
at 8 weeks of age. Hamsters were killed at 78 weeks of age.
When dietary fat was low (11% calories as corn oil), the low-
selenium group had 25 pancreatic ductular carcinomas (PCDA) in
18 hamsters, and the high-selenium group had 63 PCDA in 23
hamsters. When dietary fat was high (45% of calories as corn
oil), the low-selenium group had 27 PCDA in 19 hamsters and
the high-selenium group had 44 PCDA in 18 hamsters.

Other studies have been reported. Milner, who showed that
pharmacological doses of selenium inhibited the growth of
transplantable tumours, has recently summarized his work
(Milner, 1985), and Thompson has summarized his work on
mammary carcinogenesis (Thompson, 1984). The metabolism of
the carcinogen 2-acetylaminofluorene in selenium-deficient

rats has been studied by Besbris et al. (1982). They found
that selenium-deficient rats excreted more N-OH-acetylamino-
fluorene than controls, suggesting that selenium might prevent
the production of this carcinogenic metabolite or promote its
detoxification.

8. EFFECTS OF SELENIUM ON MAN

8.1 High Selenium Intake

8.1.1 General population

8.1.1.1 Signs and symptoms

When it became apparent that selenium was the toxic factor in plants that caused alkali disease in livestock raised in seleniferous areas, public health personnel became interested in the possible hazards for human health in such regions, since seleniferous grains or vegetables grown on high-selenium soil could enter the human food chain. Smith et al. (1936) reasoned that, if human selenosis were a problem anywhere, it would most likely occur in farmers living in seleniferous regions, who consumed largely locally-produced foodstuffs. Therefore, they surveyed a rural population living on farms or ranches known to have a history of alkali disease. Their survey inquired into the health status of 111 families and also determined the actual consumption of locally-produced foods. Wherever possible, general physical examinations were made and urine samples were collected.

These workers were unable to find any symptom or group of symptoms or serious illness that could be considered characteristic of, or could definitely be attributed to, selenium poisoning in man. However, the incidence of vague symptoms of ill health and symptoms suggesting damage to the liver, kidneys, skin, and joints was rather high. But since the causes for such disorders are many, it was not possible to determine whether selenium played any role in their causation. Apart from the more vague symptoms of anorexia, indigestion, general pallor, and malnutrition, the following more pronounced disease states were observed: bad teeth, yellowish discoloration of the skin, skin eruptions, chronic arthritis, diseased nails, and subcutaneous oedema.

The same authors (Smith et al., 1936) were unable to demonstrate a very definite correlation between the clinical evidence of selenium intoxication in 127 subjects and the concentration of selenium in the urine. In fact, they expressed surprise that there was not any more definite evidence of serious injury, particularly in subjects with high concentrations of selenium in the urine. Relatively high urinary-selenium levels were most often associated with pathological nails, gastrointestinal disorders, icteroid skin, and bad teeth. The incidence of high urinary-selenium in individuals with dermatitis and arthritis was no greater than that in individuals without symptoms.

Because of the ambiguous findings in their first study, Smith & Westfall (1937) carried out a second, more detailed and intensive survey to establish the symptomatology of human selenosis and its relation to the amount of selenium excreted in the urine. For this purpose, they examined 100 subjects from 50 families that had high levels of urinary-selenium in their previous testing. The percentage frequency of the various signs and symptoms observed is given as follows: none, 24; gastrointestinal disturbances, 31; icteroid discolouration of the skin, 28; bad teeth, 27; sallow and pallid colour, especially in younger individuals, 17; history of recurrent jaundice, 5; dermatitis, 5; pigmentation of the skin (chloasma?), 3; pathological nails, 3; rheumatoid arthritis, 3; cardiorenal disease, 2; vitiligo, 2. It was concluded that none of these signs or symptoms could be regarded as specific for selenium poisoning, and it was not certain that any one was the direct result of the continual ingestion of selenium. Nonetheless, these workers felt that the high incidence of gastrointestinal disturbances was of importance. Moreover, the high incidence of icteroid discolouration of the skin was thought to be related in some way to the ingestion of selenium, possibly as a result of liver dysfunction. The possible cariogenic effects of high levels of selenium is discussed further in section 8.1.1.2. The other symptoms occurred so infrequently that they did not seem to be associated with selenium. However, it should be noted that the value of all these observations is in doubt since there is no comparative information available concerning the frequency of occurrence of these signs and symptoms in an appropriate control group.

Smith & Westfall (1937) estimated that most of their subjects living in highly seleniferous areas were probably absorbing about 10 - 100 µg/kg body weight per day and that some of their subjects might have absorbed as much as 200 µg/kg per day. For a 70-kg man, these rates of absorption would be equivalent to a dietary-selenium intake of 700 - 14 000 µg/day, assuming that all of the selenium in the food had been absorbed.

As discussed in section 4.1.2, drinking-water rarely contributes much selenium to a person's total daily intake, but a brief note indicated that a family of North American Indians living in Colorado, USA, suffered hair loss, weakened nails, and listlessness after consuming well water reported to contain 9000 µg selenium/litre for about 3 months (Anonymous, 1962). This incident was thought to be the first authentic case of selenium poisoning in human beings induced exclusively from a naturally-occurring underground source of water. Tsongas & Ferguson (1977) studied the effects of selenium in the drinking-water on the health of 2 groups of persons living in a rural Colorado community. The first group ("exposed") received a water supply that contained 50 - 125 µg/litre, whereas the second group ("unexposed") received a water supply that contained levels ranging from non-detectable (1 µg/litre) to 16 µg/litre. Examination of a total of 86 individuals revealed that there were no significant differences between the exposed and unexposed groups in the incidence or prevalence of any of 85 health variables studied.

Lemley (1940) presented a 58-year-old rancher from South Dakota, thought to be the first described case of chronic selenium dermatitis in a human being, caused by the ingestion of selenium from natural sources. However, 2 samples of urine collected from this patient a week apart, contained only 0.043 and 0.040 mg selenium/litre, levels that are considered to be within the normal range. Although some improvement in the patient's condition was noted when certain seleniferous foods consumed on his ranch were avoided, the improvement was not marked. Administration of bromobenzene, now known to be a potent hepatotoxin, cleared up the dermatitis, but caused only a small increase in the urinary-selenium output, which reached a peak value of about 0.100 mg/litre, after 4 days. Earlier work by Moxon et al. (1940) had shown that the rate of excretion of selenium from selenized animals could be increased by the administration of bromobenzene. The likelihood that selenium was the cause of the dermatitis in this patient is doubtful, since, in a later paper, Lemley & Merryman (1941) claimed that a relapse in this patient's condition was cured when some canned meat containing 0.40 µg selenium/kg (considered by them to be a highly toxic amount of selenium) was withdrawn from the patient's diet. This figure for the level of selenium in the meat is almost certainly erroneous, since not only is it below that which would be found in animals suffering from selenium deficiency but it is also below the sensitivity of the analytical techniques available at that time.

Other cases of presumed human overexposure to selenium were reported by Lemley & Merryman (1941). For example, urine samples from a ranching family living in South Dakota contained 0.200, 0.250, 0.300, 0.550, and 0.600 mg selenium/litre for the father, mother, daughter, son, and uncle, respectively. Administration of bromobenzene to the father resulted in a urinary output of selenium of 1.800 mg/litre within 24 h. No mention of dermatitis was made in connection with this family. Rather, all family members suffered from slight, continual dizziness and clouding of the sensorium, extreme lassitude accompanied by depression, and moderate emotional instability. The patients tired on exertion and complained that their powers of concentration were markedly impaired. After a course of bromobenzene, these people improved markedly and their general condition remained improved, since they were instructed not to use the various food products of the ranch that contained selenium. Lemley & Merryman concluded that the following facts provided the basis for the diagnosis of selenium poisoning in these subjects:

(a) knowledge that the subjects lived in a seleniferous area;

(b) presence of concentrations of selenium in the urine exceeding 0.100 mg/litre;

(c) increased elimination of selenium in the urine after bromobenzene administration; and

(d) improvement in the subject's symptoms after elimination of selenium from the diet.

The same authors also described a 65-year-old rancher from South Dakota, who suffered alternate bouts of diarrhoea and constipation. This patient's urinary-selenium concentrations were as high as 0.250 mg/litre. An exploratory abdominal operation suggested that the patient had very early cirrhosis of the liver, which was confirmed by pathological examination. The patient recovered from the operation and then gradually improved after being placed on a strict selenium-free diet and given a course of bromobenzene.

In a later paper, Lemley (1943) expressed the opinion that human selenium poisoning is common, widespread, and, in certain localities, of importance for the general public health. More recently, Kilness (1973) complained that no follow-up public heath survey with an appropriate control group had been made in South Dakota since the initial surveys by the United States Public Health Service in highly seleniferous areas more than 30 years previously.

Jaffe (1976) carried out a field study in Venezuela and compared 111 children living in a seleniferous area (Villa Bruzual) with 50 living in Caracas. The overall haemoglobin and haematocrit values were somewhat lower in Villa Bruzual (128 g/litre and 39 volume %, respectively) than in Caracas (148 g/litre and 42 volume %, respectively), but no correlations between blood- and urine-selenium levels and haemoglobin or haematocrit values were found. The children in Villa Bruzual consumed less meat and milk and had a higher incidence of intestinal parasite infestation (Jaffe et al., 1972a) than those in Caracas. Therefore, it was concluded that the differences in haemoglobin were probably due to differences in nutritional or parasitological status and not to differences in selenium intake. Activities of prothrombin and serum alkaline phosphatase and transaminases, which altered in selenium-poisoned rats (Jaffe et al., 1972b), were normal in all the children and no correlation with blood-selenium levels was apparent. Symptoms of dermatitis, loose hair, and pathological nails were reported as more frequent among the children in the seleniferous area than in those living in Caracas, but no quantitative information was given. The clinical signs of nausea and pathological nails appeared to be correlated with serum- and urine-selenium levels. However, the cause of the different incidence of clinical signs was considered doubtful, especially in the absence of differences in the various biochemical tests performed (Jaffe et al., 1972a).

The mean blood-selenium level of 111 school children from the seleniferous area in Venezuela was 0.813 mg/litre (Jaffe et al., 1972a). A subgroup of 28 children from this area who had blood-selenium levels of over 1 mg/litre had a mean blood-selenium level of more than 1.321 mg/litre. There was one child with a blood-selenium level of 1.8 mg/litre. Another subgroup of 11 children who had blood-selenium levels of less than 0.4 mg/litre had a mean blood-selenium level of 0.330 mg/litre. Urinary-selenium levels tended to reflect blood-selenium levels, since children in the high-blood-selenium subgroup (> 1 mg/litre) excreted a mean level of 0.657 mg/litre urine, whereas children in the low-blood-selenium subgroup (< 0.4 mg/litre) excreted a mean level of 0.266 mg/litre urine.

Kerdel-Vegas (1966) summarized 9 cases of acute intoxication due to the ingestion of nuts of the "Coco de Mono" tree (Lecythis ollaria) from the seleniferous areas of Venezuela. Although the course of the poisoning differed from patient to patient (probably because of differences in the amount of nuts consumed and the length of time the nuts were present in the digestive tract), most cases experienced nausea, vomiting, and diarrhoea, a few hours after eating the nuts, followed by hair loss and nail changes, some weeks after the initial episode. Two patients had foul breath that was described as being reminiscent of decomposed seaweed or phosphorus. After a period of time, re-growth of hair and nails took place and most patients appeared to make a satisfactory recovery. One fatality was reported, a 2-year-old boy who died, in spite of proper medical attention and symptomatic treatment for severe dehydration.

Dickson (1969) described his own personal experience after eating sapucaia nuts (Lecythis elliptica H.B.K.) that had grown in Honduras. He reported hair loss and splitting of finger and toenails several weeks after consuming the nuts. Although no mention of selenium was made in his report, he noted that a peculiar odour was associated with consumption of these nuts, not only of his breath but of his body as well. Kerdel-Vegas (1966) pointed out that the sapucaia, or chestnut of Para (which he called <u>Lecythis paraensis</u> Ducke), is consumed as food in Brazil and other countries to which it is exported (i.e., Europe and the USA). He also commented that there is a popular saying in northern Brazil that eating sapucaia leads to hair loss, but he knew of no scientific or medical reports from Brazil. Signs of selenium toxicity have been observed in rats fed diets containing defatted Brazil nut flour that assayed 51 mg selenium/kg (Chavez, 1966), and Thorn et al. (1978) found that samples of Brazil nuts marketed in the United Kingdom contained an average of 22 mg selenium/kg (range, 2.3 - 53 mg/kg).

As reported recently, an unexplained intoxication characterized by nail deformation and the loss of hair and nails as the most common signs was observed in Enshi county of Hubei province of the People's Republic of China, more than 20 years ago, with a peak prevalence during the years 1961-64 (Yang et al., 1983). In 5 villages with 248 inhabitants, about half of the population was affected and in one particular village over 80% of the people were affected. No other quantitative information about the cases was given but a general description of the signs and symptoms observed is presented below. The hair became dry and brittle and was easily broken off at the scalp with retention of intact radicles so that depigmented and dull hair continued to grow. A scalp rash accompanied by intolerable itching resulted in hard scratching which easily removed the hair. The nails became brittle with white spots and longitudinal streaks on the surface, followed by a break on the wall of the nail. With new nail growth, the broken nail advanced and ultimately fell off. Effusian of fluid from around the nail was common in many cases. The new nail had a rough and ridged surface and was fragile and thickened.

Other signs of intoxication included skin lesions, tooth decay, and abnormalities of the nervous system. The skin became red and swollen and then blistered and eruptive followed, in some cases, by ulcerations that took a long but unstated time to heal. The lesions occurred primarily on the limbs and also on the back of the neck. Mottled teeth were observed in the intoxicated individuals but this observation may have been confounded by high exposure to fluoride. In one heavily affected village, nervous system abnormalities were seen, including peripheral anaesthesia, acroparaesthesia, and pain in the extremities. Hyperreflexia of the tendon commonly developed later followed by numbness, paralysis, and motor disturbances. One case of hemiplegia was also reported. There was an indication that this intoxication was related to the locally grown monotonous diet consumed, which later was shown to contain high levels of selenium. No quantitative data regarding selenium exposure were obtained at the time of the outbreak. However, because of circumstantial evidence, the intoxication was considered by the authors to be selenosis, and, as discussed in sections 5.1.1.1 and 6.2.3, current levels of exposure to dietary selenium in an area with a history of intoxication, as assessed by blood- and hair-selenium levels as well as dietary intake data, exceeded those ever reported in any non-occupationally exposed population (Table 8). The authors estimated that the daily dietary intake of selenium, after the peak prevalence of the poisoning had subsided, averaged 5 mg, and blood- and hair-selenium levels averaged 3.2 mg/litre and 32.2 mg/kg, respectively. The ultimate environmental source of the selenium in this episode of intoxication was a highly seleniferous coal (average selenium content of 300 µg/g) which lost its selenium to the soil as a result of weathering processes. Once in the soil, the selenium was taken up by plant crops and entered into the food chain.

Because of the nutritionally-beneficial effects of selenium in animals, some investigators have deliberately given inorganic or organic forms of the element to people with the aim of producing some desirable health benefit. For example, Westermarck (1977) administered selenium, as selenite, in oral doses of 0.05 mg/kg body weight per day, for more than one year, to patients with neuronal ceroid lipofuscinosis (NCL) and did not observe any toxic manifestations. On the contrary, it was felt that some of the NCL patients showed at least a transitory improvement in their condition. In some patients, a slight increase in serum aspartate aminotransferase activity was observed, but apparently this is seen in a number of patients with NCL.

Schrauzer & White (1978) described 2 individuals who had been taking commercially-available nutritional supplements consisting of selenium-containing yeast at doses of 200 and 450 µg selenium, daily, for 18 months. Together with their dietary intakes, these individuals received a total of 350 and 600 µg/day. Although some marginal haematological changes were seen and the serum-glutamic oxaloacetic transaminase activities were on the borderline of high, it was concluded that daily intakes of up to 600 µg selenium for 18 months do not induce toxic effects in well-fed individuals.

In a study by Perona et al. (1978), 4 subjects were given, orally, a 2-mg dose of selenium, as sodium selenite, daily, for 20 - 40 days, to determine the effects of in vivo selenium administration on the activity of human erythrocyte-glutathione peroxidase. It was stated that none of the subjects exhibited any symptoms of selenium poisoning, but the criteria used to judge any deleterious effects of selenium were not described.

Yang et al. (1983) reported the case of a 62-year-old man who had taken one tablet containing 2 mg sodium selenite per day for more than 2 years. The subject did not have any symptoms of indisposition but presented with thickened, fragile and somewhat honeycomb-like fignernails. After the oral intake of sodium selenite was stopped, the surface of the new nail growth became smooth and gradually recovered. Blood- and hair-selenium levels were 0.179 mg/litre and 0.828 mg/kg, respectively, on the day the selenite tablets were discontinued. These tissue-selenium levels are much lower than those reported by the same authors in Enshi county of the Hubei province (Yang et al., 1983) suggesting that the form of selenium ingested must be considered when interpreting tissue-selenium levels in the diagnosis of selenium poisoning.

Ingestion of superpotent selenium tablets, meant to be consumed as a "health food" supplement, resulted in 12 cases of human selenium toxicity in the USA in 1984 (Anonymous, 1984; Jensen et al., 1984; Helzlsouer et al., 1985). Each tablet contained 27 - 31 mg selenium by analysis (about 182 times more than the level stated on the label). Approximately 25 mg of the selenium was present as sodium selenite whereas the rest was elemental and/or organic selenium. The total doses of selenium estimated to be consumed by the victims ranged from 27 to 2387 mg. Based on the limited information available, the symptoms reported in these cases as most common were nausea and vomiting, nail changes, hair loss, fatigue, and irritability. Other symptoms included abdominal cramps, watery diarrhea, paraesthesias, dryness of hair, and garlicky breath. Eight of the 12 victims did not have any abnormalities in the blood chemistry, and the results of liver and kidney function tests were normal.

A level of 500 µg has been proposed as the tentative maximum acceptable daily intake of selenium for the protection of human health (Sakurai & Tsuchiya, 1975). This figure was derived from an initial estimate of the mean normal daily intake of selenium by human beings of between 50 and 150 µg. It was concluded that values of 10 - 200 times the normal intake appeared acceptable as an estimated range for the margin of safety within which the average human being could tolerate selenium. By taking the lower values of both these estimates, the lowest level of potentially dangerous daily intake of selenium was estimated to be 500 µg.

8.1.1.2 <u>Attempts to associate high selenium intake with human diseases</u>

(a) <u>Dental caries</u>

As already mentioned in section 8.1.1.1, bad teeth was one of the signs that was thought to be possibly associated with a high intake of selenium by a rural population living in seleniferous areas of South Dakota, Wyoming, and Nebraska (Smith et al., 1936) and by children living in a seleniferous zone in Venezuela (Jaffe, 1976). Several studies have been concerned with the possibility of a specific association between the incidence or prevalence of dental caries and residence in a seleniferous area, urinary excretion of selenium, or the selenium content of teeth. Hadjimarkos & Storvick (1950) and Hadjimarkos (1956) noted a geographical difference in the prevalence of dental caries in 2029 children, between the ages 14 and 16 years, residing in 4 different counties in Oregon. Two different counties, one with the highest (Clatsop), and one with the lowest (Klamath), rates of caries experience were selected for further study. Urine samples from 24 and 29 male children who were attending county seat high schools and who were born and residing in the respective counties were analysed for selenium content. The mean levels of urinary-selenium were 0.049 and 0.037 mg/litre in the groups of children from the counties with the high (14.4 DMF (decayed, missing, or filled) teeth per child) and low (9.0 DMF teeth per child) prevalence of dental caries, respectively (Hadjimarkos et al., 1952).

In a second study, 2 additional counties with similar but high rates of caries experience (Jackson and Josephine, 13.4 and 14.4 DMF teeth per child, respectively) were selected (Hadjimarkos & Bonhorst, 1958). Analysis of 33 urine specimens from continuously-resident, male high school children attending a county seat high school (Jackson) and 46 specimens from similar children attending a high school serving a rural area (Josephine) gave values (mean ± SE) of 0.074 ± 0.007 and 0.076 ± 0.005 mg/litre, respectively. It was concluded that there was an association between the high prevalence of dental caries and the values of urinary-selenium excretion, which were double those seen in the previous study in the county with a lower rate of caries experience.

In order to explain the variations in urinary-selenium levels observed in the above studies, samples of locally-produced and -consumed milk and eggs, as well as local drinking-water samples were collected from 74 farms located in Klamath, Jackson, and Josephine counties and analysed for their selenium content (Hadjimarkos & Bonhorst, 1961). The selenium levels were almost 10 times higher in food samples collected from the counties with high caries prevalence (Jackson and Josephine) than in those from the county with low caries prevalance (Klamath). The selenium levels in the milk samples obtained in Jackson and Josephine counties were higher than those listed in Table 3.

In another study, Tank & Storvick (1960) determined the extent of dental caries in a population of children from 15 areas of Wyoming that had been identified previously as seleniferous or non-seleniferous on the basis of the geological distribution of selenium, occurrence of the element in vegetation, and the occurrence of selenosis in livestock. These research workers claimed that the dental caries rate in the permanent teeth was higher in children residing in seleniferous areas of Wyoming than in children living in non-seleniferous areas of Wyoming, but they were unable to demonstrate any consistent relationship between calculated selenium intake and caries or urinary-selenium excretion and caries.

Suchkov et al. (1973) found that the incidence of caries varied in 3 different geographical zones (mountainous, pre-mountainous, and forest-steppe) in the Chernovitsi region of the Ukraine. Analysis of teeth from people residing in rural areas in these 3 zones and mainly consuming locally-produced foodstuffs revealed a direct correlation between the selenium content of the teeth and the incidence of dental caries. The people in the mountainous zone had the highest level of selenium in the teeth and the greatest incidence of dental caries, whereas the people in the forest-steppe zone had the lowest level of selenium in the teeth and smallest incidence of dental caries. This correlation was true for deciduous teeth as well as for permanent teeth. However, as pointed out by the authors, the mountainous zone also had the softest water and the lowest content of fluorine in the drinking-water, so that factors other than selenium might have been involved.

It was not possible to demonstrate any association between the urinary excretion of selenium and the incidence of dental caries in subjects living on the South Island of New Zealand, an area known to be low in selenium (Cadell & Cousins, 1960). However, Hadjimarkos (1960) claimed that the levels of urinary excretion of selenium in this study (all less than 0.050 mg/litre) were too low to be associated with an increased incidence of caries.

As emphasized by Schwarz (1967), the levels of urinary-selenium excretion reported in the Hadjimarkos studies were within the limits generally found in the normal population without excessive selenium intake. It may be recalled from section 7 that, in experimental animal studies, very high toxic levels were used to demonstrate the cariogenic effect of selenium.

(b) Reproduction

Possible effects of selenium on reproduction were sus-pected following old anecdotal reports from Colombia, South America that women living in areas, later shown to be seleniferous, gave birth to malformed babies (Rosenfeld & Beath, 1964), but there is a lack of reliable studies. Robertson (1970) pointed out that, among 6 women preparing microbiological media containing sodium selenite, one probable and 4 certain pregnancies all ended in abortions, save one that went to term. The infant was born with bilateral club foot. However, no differences in urinary-selenium levels were observed between this group of women and a control group living in the same area, and inquiries by the author at other laboratories carrying out comparable work did not show any evidence of similar trouble. Jaffe & Velez (1973) could not demonstrate any correlation between the selenium level in the urine of school children in different Venezuelan states and the incidence of infant mortality due to congenital malform-ations, on the basis of published public health statistics. Jaffe (1973) concluded that no recent observations of a teratogenic action of dietary selenium in human beings had been reported at that time.

(c) Amyotrophic lateral sclerosis

Kilness & Hochberg (1977) reported an unusual cluster of 4 cases of amyotrophic lateral sclerosis (ALS) in male farmers living in a seleniferous area and indicated that selenium might be an environmental factor predisposing to the disease. But Schwarz (1977) pointed out that the frequency of ALS was at least as high if not higher in areas that were selenium-deficient as in those with normal or elevated levels and Kurland (1977) suggested that the cluster was more indicative of a chance occurrence than of a new etiological lead in ALS. Moreover, Norris & Sang (1978) found that 19 out of 20 well-established cases of ALS had urinary-selenium levels lower than the mean for unexposed persons and therefore concluded that selenium exposure was of no concern in the average case.

8.1.2 Reports on health effects associated with occupa-tional exposure

Although the toxicological potential of selenium for human beings can be inferrèd from studies carried out with labora-tory animals, certain precautions must be taken in applying

the results of animal studies to the industrial health
aspects of selenium. First, the number of studies dealing
with the respiratory exposure of animals to selenium compounds
is limited, and the dose levels employed have usually been
higher than those encountered in industry. On the other hand,
the period of exposure used in the studies reviewed in section
7.1.2.3 did not exceed one month. The specific chemical form
and physical state of the selenium must be considered as well
as the fact that the chemical form of selenium may change when
in contact with moist mucous membranes or with sweat. These
factors, plus several others, must be taken into account when
considering the health effects of various selenium compounds
under occupational exposure conditions.

The Task Group recognized that, for various reasons, knowledge on the health effects of industrial exposure to selenium
compounds was not complete. Acute exposures to selenium are
the result of accidents and are, of necessity, described on an
ad hoc case study basis. Regarding the effects of long-term
selenium exposure in industry, there is a lack of epidemiological studies that include unexposed control groups. Also,
no follow-up studies are available comparing the health status
of a sufficiently large number of workers, previously exposed
to selenium, with that of the unexposed population. Furthermore, the exposure level was not known with certainty in
either acute or long-term studies, and, in some cases, the
form of selenium was not established. In many cases, simultaneous exposure to other noxious agents occurred. Nevertheless, the Task Group felt that some preliminary assessment
of the toxicological potential of selenium in industry could
be made on the basis of the occupational experience available
with selenium compounds.

Since Hamilton's first observation on the effects of
selenium exposure in industry in 1917 (Hamilton, 1927-34),
several hundred persons have been described in the literature
who have been directly affected by the vapour, fumes, or dust
of selenium and its compounds. In addition, there is a systematic study of a group of over 100 selenium workers for a
period of 2 years (Glover, 1967). Systematic attention has
also been given to the health of workers exposed to selenium
in the USSR (Filatova, 1948; Monaenkova & Glotova, 1963; Gracianskaja & Kovshilo, 1977). Several reviews evaluating the
health effects of selenium from the point of view of occupational health have been presented (Glover, 1954-70, 1976;
Cooper, 1967; Izraelson et al., 1973; Cooper & Glover, 1974).

8.1.2.1 Fumes and dust of selenium and its compounds

Workers in several industries, such as the production or
recovery of selenium itself, or the manufacture of glass or

rectifiers can be exposed to the fumes and dust of selenium and its compounds. In addition, direct contact with powders or solutions containing selenium compounds is possible and the biological effects of such contact will be discussed in this section.

The chemical form of selenium in the fumes and dusts found in the above-mentioned industries, consists of elemental selenium plus various amounts of selenium dioxide, the only oxide of selenium found in the industrial environment. However, in some cases, the possible occurrence of other selenium compounds, such as hydrogen selenide, cannot be excluded. Instances where hydrogen selenide was the main selenium compound of concern will be discussed separately in section 8.1.2.1.2.

8.1.2.1.1 Selenium dioxide

Selenium dioxide mainly occurs in industry:

(a) during the production of the compound, usually by the oxidation of elemental selenium; and

(b) whenever selenium is heated in air, purposely or accidently, above its melting point. "Selenium fume" rising under these conditions is a mixture containing red elemental selenium and about 20 - 80% of selenium dioxide.

Glover (1954, 1970, 1976) has summarized the acute local, systemic, and possible long-term effects of occupational exposure to selenium dioxide. Acute local effects are seen in the lungs, gastric mucosa, skin, nails, and eyes. Sudden inhalation of large amounts of selenium dioxide produces pulmonary oedema, due to a local irritant effect on lung alveoli. Working in an atmosphere containing selenium dioxide can increase indigestion. Contact with the skin results in burns or dermatitis. Occasionally, a true urticarial type of generalized allergic body rash may occur, in which case the individual must be removed from any work involving selenium. The selenium dioxide may penetrate under the nails and cause excruciating pain in the nail beds. If selenium dioxide enters the eyes, prompt flushing with water will prevent conjunctivitis. "Rose eye", a pink discoloration of the skin of the eyelids, which often become puffy, is sometimes seen in persons who work in an atmosphere of selenium dioxide dust. Systemic effects of selenium dioxide exposure include garlicky-smelling breath, metallic taste on the tongue, and indefinite sociopsychological effects. The garlicky breath is the first and most characteristic sign and has been used by occupational hygienists as a way of monitoring exposure, even though the odour is not always a reliable index in this respect. The metallic taste is an earlier symptom but, being more subtle, is often overlooked by workers. Sociopsychological effects such as lassitude and irritability have also been associated with selenium exposure. Since the effects of selenium overexposure lead to hepatic damage in animals, Glover (1954) concluded that it would be prudent to watch for such damage in human beings.

(a) Reports on health effects connected with short-term accidental exposures

An industrial incident in a plant engaged in smelting scrap aluminium and other non-ferrous metals was reported by Clinton (1947). The aluminium scrap included more than 100 kg of aluminium rectifier plates coated on one side with metallic selenium overlaid with a coating of an alloy of bismuth, cadmium, and tin. When an attempt was made to skim the dross prior to pouring, a cloud of reddish fume arose. The fumes were intensely irritating to the eyes, nose, and throat and the plant was evacuated. No workers were exposed for more than 2 min.

All exposed workers noticed immediate and intense irritation of the eyes, nose, and throat, and an unpleasant sour, garlic-like odour to the fumes. The more heavily exposed workers complained of a severe burning sensation in the nostrils, and dryness of the throat; followed after 2 - 4 h by severe headache, mainly frontal in location, lasting until the following day. Several men observed immediate sneezing, coughing, and headache, followed for 4 - 8 h by nasal congestion, dizziness, and redness of the eyes. Most of the men noticed a bad taste in their mouths and an unpleasant odour to their skin and clothing. Other people, however, did not note any unpleasant odour in the breath of the exposed workers.

Two labourers who had attempted to skim the dross from the furnace and therefore underwent intense exposure were hospitalized for observation. On arrival at the hospital, they complained of soreness of the eyes and lachrymation, pain in the nose, slight difficulty in breathing, and frontal headache. Physical examination on admission revealed conjunctival injection, congestion of the mucous membranes of the nose and throat, and oedema of the uvula. One man had a few fine rales audible in the right base; roentgenological examination of the chest did not reveal any abnormalities. The temperature, pulse, and respiration were normal. Both workers were discharged 2 days after the accident, at which time they were asymptomatic and had no positive physical findings. Another worker, a foreman, underwent a more severe exposure than most of the workers. He was not hospitalized, but, about 8 - 12 h after the accident, he developed severe dyspnoea accompanied by a slight elevation in temperature, in addition to a headache and sore throat. Physical examination revealed fine rales in both bases, as well as scattered asthmatic-like wheezes throughout the chest. These signs and symptoms cleared in about 24 h, without specific therapy. All persons exposed to the fumes had recovered entirely in 3 days, and no sequelae have been encountered.

In the same year, Lauer (1947) described an episode that occurred when 15 men were exposed to fumes arising from an explosion and fire, which occurred in a pot when selenium became overheated in contact with aluminium. The 2 elements produced a high-temperature reaction, and fumes and vapours were dispersed throughout the work area. The lengths of exposure varied, some victims were using masks part of the time, and others were almost without protection.

The exposed men reported to the medical dispensary about 15 - 45 min after the accident. All complained of soreness and burning of the nose and throat, and 8 out of 15 cases had some dyspnoea. Two of these cases were moderately severe, requiring almost continuous oxygen therapy. Headache and dizziness and a burning sensation in the eyes occurred in 6 cases. Three others complained of substernal burning and tightness of the chest. In 4 there was nausea and vomiting. All were hospitalized. Physical examination showed reddening of nasal and pharyngeal mucosa, wheezes, and musical rales in the lungs (12 cases). These symptoms abated quickly. In about 10 days, there were few subjective complaints. All apparently recovered satisfactorily with no known complications. To assess possible liver affection, Hanger's cephalin cholesterol tests and Icterus Index determinations were performed on admission to the hospital and again at the end of the first, second, third, and seventh week. On the day of admission to the hospital, Hanger's tests were made on 13 workers, 4 of which were positive and 9, negative. At the end of the first week following exposure, there were 10 positives and 3 negatives. At the end of the second week, there were 11 positives and 3 negatives out of a group of 14 tested, and the same result was obtained at the end of the third week. By the end of the seventh week, only 4 of 11 tested were positive. The average value of the Icterus Index on admission to the hospital was 7.3 units. One week after exposure, it was 13.5. In 2 weeks, it averaged 20.7. After the third week, it had fallen to 9.9 and, by the end of the seventh week, the average Icterus Index was 7.2.

An accident connected with a fire in a selenium rectifier plant was described by Wilson (1962). Twenty-eight of the employees were exposed directly to the smoke and fumes containing selenium dioxide. The length of exposure to the fumes varied with the individual, but no one was exposed for more than 20 min. Oxygen was administered by mask to the men lying on the ground, who experienced bronchial spasm with coughing, gagging, and, in some instances, transient loss of consciousness.

The initial signs and symptoms were a feeling of constriction in the chest, accompanied by burning and irritation of the upper respiratory passages, violent coughing and gagging with nausea and vomiting, and a bitter acid taste in the mouth. During the acute episode, there were mild signs of shock, with a drop in blood pressure and an elevated pulse and respiratory rate. Other symptoms, experienced during this stage, were burning of the skin, conjunctivae, and mucous membranes of the upper respiratory passages. Within 4 h, all patients had apparently recovered from the acute episode and a recheck of their blood pressure, pulse, respiratory rate, and general condition was found to be within normal limits.

Within 6 h of the exposure, the victims began complaining of secondary symptoms with the onset of generalized chills accompanied by nausea and vomiting, diarrhoea, malaise, dyspnoea, and headache. The onset of secondary symptoms was delayed in some of the victims for several hours after the initial exposure. Within 12 h, all exposed personnel began experiencing the symptoms described. The following morning, the first patient, a 30-year-old male, was admitted to the hospital. He was cyanotic and in moderate respiratory distress. He complained of chest pain bilaterally, and was experiencing bronchial spasms. A chest roentgenogram revealed extensive bilateral consolidation of the lung fields, indicating pneumonia. The white blood cell count was elevated to 153 000 with a marked prominence of neutrophils. He experienced a stormy course, requiring constant oxygen for the first 9 days of hospitalization. By the fifth hospital day, X-ray films revealed a further extension of the previously noted bilateral pneumonic consolidation. Two weeks following admission, comparison of a chest film with previous films revealed marked improvement bilaterally. He was free of respiratory distress, his white blood cell count had returned to within normal limits, and he was discharged home for convalescence. It is interesting to note that this was the only employee of the 5 hospitalized cases who did not receive oxygen on the afternoon of the fire. On the same day, the second victim was admitted to the hospital in a similar dyspnoeic, cyanotic state with an elevated white blood cell count of 24 600. X-ray again revealed bilateral atelectasis and consolidation. This patient required continuous oxygen for 6 days, and was discharged after 9 days of hospitalization: his white blood cell count had returned to normal and a repeat roentgenogram revealed improvement. This was the second most involved and prolonged illness, and he received only about 6 breaths of oxygen following his exposure. Three days after the accident, a third employee was admitted to the hospital in less respiratory distress, with a normal white blood cell count and a chest film that revealed bilateral elevation of the diaphram with extensive peribronchial infiltration and some consolidation at the lung bases. On the same day, another worker was admitted to the hospital with a normal white blood cell count and a chest film revealing bilateral increase in lung markings consistent with bronchitis; however, no consolidation was evident. The final patient to be hospitalized was admitted 4 days after the accident with a white cell count of 12 800 and a prominence of neutrophils. The X-ray film indicated bilateral pneumonia.

Four days after exposure, all other employees with any upper respiratory complaints were examined. Thirty-two of the 53 employees examined were found to have a residual bronchitis, of minimal to moderate degree, that required medication. These were treated on an outpatient basis. Within 1 week, all were asymptomatic and free of any respiratory signs.

Both Monaenkova & Glotova (1963) and Skornjakova et al. (1969) have described occupational incidents of short-term selenium dioxide exposure. In the former case, 6 workers, 20 - 52 years old, were exposed from 5 to 20 min in a room with a leaky selenium still. The workers suffered acute pain in the eyes, hoarseness in the throat, painful cough, and a heavy feeling in the chest. Acute conjuctivitis, laryngotracheitis, and bronchitis were also noted and gastroenterocolitis was common to all. Transient fever was seen in 2 workers. The various signs and symptoms disappeared within 5 - 7 days, but, on return to work involving selenium, 2 workers developed chronic bronchitis. In the report of acute selenium dioxide exposure by Skornjakova et al. (1969), there was irritation of the eyes and mucous membranes, coughing without expectoration, nasal secretion, headache, vertigo, nausea, vomiting, garlicky-smelling breath and skin, general weakness, loss of consciousness, and collapse. Two weeks after exposure, an allergic whole body rash occurred that disappeared after removal of the worker from contact with selenium.

(b) Effects of short-term and/or repeated dermal exposure

Selenium dioxide is more than a primary irritant, it causes extremely painful burns on the skin which, however, always heal without a scar. Theoretically, the selenium dioxide powder itself does not burn the skin (and if dropped on to the skin should be immediately brushed off dry).

However, in practice, in the industrial environment, there is sufficient moisture on the skin from sweating for this very deliquescent white solid to form a sticky solution of selenious acid within seconds, or at the most, minutes, of coming into contact with the skin. Selenious acid is so soluble that, if the skin is immediately washed with a lot of water, there will be no burn or even rash from accidental contact. When the selenious acid does burn the skin, there is little to see or feel for several hours. After 4 h with a 50% solution, an unremitting intense pain begins, small petechial areas occur, and a faintly orange coloration occurs, which indicates reduction of some of the selenious acid to elemental red selenium. The pain and necrosis can be prevented by the application of a reducing solution or ointment, such as 10% sodium thiosulfate (Glover, 1954). If the burn remains untreated it may go on to ulceration, but this is rare. Högger & Böhm (1944) described a case of a woman who suffered pain and reddening of the middle, ring, and little fingers of one hand when selenium dioxide crystals penetrated into the protective rubber glove that she was wearing.

The first accurate description of a case of selenium fumes causing allergic dermatitis was published by Duvoir et al. (1937). A chemical engineer, who had handled selenium for 3 years in the past, developed swelling of the face with a hard urticarial oedema of the nose and cheeks extending on the neck down to his shirt collar following a 3-day exposure to selenium vapours. There was a semi-circle of vesicles on the lower lids, but the forehead and ears escaped completely. His hands showed discrete lesions. Within 24 h, the genital oedema had disappeared, after 3 days, the rash started to go, and, by 9 days, there was only desquamation left. The 5-cm plaque, occurring 1 h after a patch test with 10% potassium selenite, showed a greater than average contact sensitivity.

Another case was described (Halter, 1938) involving a glass worker whose job was to add a mixture of red selenium and sodium selenite to the glass. This man was in daily contact with selenite powder for 36 years. He complained of an oedematous erythema of the face and neck, and several hard infiltrated raised plaques on the dorsi of the hand and fingers. On examination, his nasal and laryngeal mucosae were found to be reddened, and there was a very slight conjunctivitis. There were no changes in the skin of areas protected by clothing. His only other complaint was of headache. There were no symptoms or signs referrable to the nervous or gastrointestinal systems. Selenium was detected in the urine (no actual values were given). His liver was found to be enlarged and there was an increase of porphyrins in the urine.

Pringle (1942) has described several cases of acute dermatitis resulting from contact with dry selenium dioxide or selenium dioxide dissolved in water, and 2 cases of acutely painful paronychia after accidental contact with dry selenium dioxide. In addition, several cases of mild rash on the anterior surface of both forearms, but principally affecting the bend of the elbows, were seen among employees working with heated selenium dioxide in the fume cupboards. Only their fingers, protected at that time with gloves, entered the fume cupboard, which apparently was adequately protected. The dermatitis resembled a seborrhoeic condition and there was frequently a thin watery discharge. In the course of a year, 2 cases developed a high degree of sensitivity to selenium dioxide, so that working in the same department, though not actually on selenium, caused repeated recurrences and they had to be removed permanently to another part of the factory.

(c) Studies on the health effects of long-term exposure

Filatova (1948) reported that workers exposed over a long period of time to selenium aerosols containing elemental selenium at levels of 0.35 - 24.8 mg/m^3 and selenium dioxide at levels of 0.11 - 0.78 mg/m^3 developed rhinitis, nasal bleeding, headaches, loss of weight, irritability, and pain in the extremities (Izraelson et al., 1973).

Conditions of exposure and the related health effects in a rectifier factory were investigated by Kinnigkeit (1962). Air-selenium levels were determined in various work-places and blood-selenium levels were measured in 62 workers exposed to selenium. Urinary-selenium levels were also determined in 22 workers. The reported air-selenium levels did not exceed 0.05 mg/m^3 and were less than the threshold limit value. However, as discussed in section 5.2.1 there was a clear-cut discrepancy between the air values and the selenium levels observed in blood and urine, indicating that the actual workers' exposure must have been considerably higher. Of 62 workers, over half (35) complained of irritability, sleeplessness, loss of appetite, and nausea. Twenty six had headaches and 3 had cramplike pains in their limbs. Clinical examination revealed irritation of the mucosa in 9 workers, with conjunctivitis and slight tracheal bronchitis. Of the 2 workers that had unavoidable skin contact with selenium, one had excematous lesions of the forearms and the other had bluish-red urticarial exanthema.

The Takata reaction and the thymol test were performed on 61 of these workers, and the results were within the normal range in all employees, with the remarkable exception of workers engaged in the electrical testing of the rectifier plates. In this group of 13 people, abnormal results suggesting impaired liver function, were obtained in 8 workers, the only employees from the plant showing this pathological reaction. As underlined by Kinnigkeit, workers carrying out this process handled the plates by hand and were frequently directly exposed to fumes containing selenium dioxide arising from the burning out of plates during electrical testing. As discussed previously, this group was characterized by a very high mean blood-selenium level (geometrical mean ± SE: 15.8 ± 11.8 mg/litre). A great inter-individual variability in selenium-blood levels existed with this and other groups, but no attempt was made to relate these individual values to the results of liver function tests. Another group of workers had blood-selenium levels that were almost as high (Table 13) (section 6.2.3), but no evidence of liver dysfunction was observed in this group. However, it cannot be assumed that the form and pathway of the exposure to selenium was the same in both groups.

Monaenkova & Glotova (1963) summed up results of clinical observations on 12 people, aged 30 - 50 years, who had been occupationally exposed to selenium for 3 - 16 years and who manifested the symptoms of chronic selenium and selenium dioxide poisoning. Ten of the 12 persons had been previously engaged in enterprises producing selenium rectifiers and were exposed to both elementary selenium and selenium dioxide. The patients complained of pain in the right hypochondrium, dyspeptic phenomena, undue fatiguability, dyspnoea, weakness, and sleeplessness. Several patients complained of cough and, in 3 of them, chronic bronchitis or moderate emphysaema were established. One patient had asthmatic bronchitis. Almost all patients had various impairments of the liver and gastrointestinal tract (toxic hepatitis in 6 patients, dyskinesis of gallbladder in 4 patients, cholecystis in 4 patients, and spastic colitis in 4 patients). The patients showed an astheno-vegetative syndrome, pigmentation of exposed areas of the skin, and, in 8 persons signs of a hyperfunction of the thyroid gland (radioiodine uptake measurements) were found. Elevation of the basal metabolic rate or hyperfunction of the thyroid gland were mentioned also briefly in a previous case report (Halter, 1938) and in a review (Holstein, 1951), respectively.

A group of selenium workers in a rectifier factory was followed by Glover (1967, 1970) in 1953-56. Every 3 months, urinary-selenium levels were determined, and the workers were asked about their health and examined for the presence of garlicky-smelling breath and skin rashes. Other routine medical checks were not performed. Workers complained of indigestion and epigastric pain; severe haematemesis (with hospitalization) occurred in one case. Several men noticed symptoms of lassitude and irritability, when on selenium work, which cleared whenever they were taken away from selenium. A strong odour of garlic on the breath was detected in most workers with selenium-urinary levels of 0.5 - 1.0 mg litre, but not below these levels. The breath odour disappeared in 7 - 10 days following the removal of workers from contact with selenium. Glover (1967, 1970) was able to trace 17 deaths, occurring within 10 years, among selenium-exposed workers in the same factory (Table 52). The author recognized that the list of deaths was not complete and that the group was too small to permit far-reaching conclusions. The length of exposure to selenium in individual cases was not given. Carcinomas arose in various organs: 2 bronchus, 1 stomach, 1 colon, 1 ovary, and 1 testis.

8.1.2.1.2 Hydrogen selenide

(a) Short-term exposure

There are few cases of occupational poisoning described in which hydrogen selenide was identified as the sole causative agent. Reported cases of occupational intoxication by hydrogen selenide resulted usually from short-term accidental exposures in the chemical laboratory or industry. Apparently, the first account of an individual requiring hospital treatment after sudden exposure to this gas was by Senf (1941), describing the effects of the accidental laboratory exposure of a chemist to hydrogen selenide. The patient noticed the characteristic smell and the gas caused her eyes to weep and her nose to run. Within a few hours, her voice became hoarse and increasing dyspnoea caused her admission to hospital. On examination she was found to have a bluish-red erythema of the skin, conjunctivitis, and injection of the nasal mucous membrane. The lung oedema, ECG myocardial changes, the dark red erythema, and porphyrinuria disappeared within 10 days. In 1941, Painter described a case of a single inhalation of hydrogen selenide. After a brief "metallic" sensation, no ill effects were felt for about 4 h. Then a copious discharge from the nasal passages began. This persisted with violent sneezing for 3 or 4 days. No ill effects were noted later. A case of pure hydrogen selenide poisoning in a works chemist, producing hydrogen selenide in order to form selenides, was described by Symanski (1950). Apparently only one inhalation of the gas was followed by a sudden feeling of constriction in the chest with coughing, tears, and burning of the nose, all of which disappeared in a few moments. The patient thought that he had lost his sense of smell. Four or 5 h later, a severe cough and dyspnoea developed. At medical examination (8 h after the accident) he was dyspnoeic and febrile. He continued to cough up blood-stained frothy sputum all night, and by morning the acute symptoms had disappeared leaving him apparently with bronchitis and an irritating

Table 52. Comparison of the certified causes of death of seventeen
selenium workers with the expected distribution by cause based on
the experience throughout England and Wales

International list number	Cause of death	Observed	Expected[b]
001 – 019	tuberculosis	0	0.4
140 – 205	malignant neoplasms	6	5.1
		–	–
330 – 334	vascular lesions	0	1.4
410 – 416	chronic rheumatic heart disease	2	0.6
		–	–
420 – 422	arteriosclerotic heart disease	3	3.8
430 – 434	"other" heart disease	0	0.2
440 – 443	hypertensive heart disease	0	0.2
444 – 447	"other" hypertensive disease	1	0.2
		–	–
450 – 456	disease of arteries	0	0.2
480 – 483	influenza	0	0.1
490 – 493	pneumonia	0	0.5
500 – 502	bronchitis	2	0.9
		–	–
–	all other causes	3	3.5
		–	–
–	All causes	17	17.1

[a] From: Glover (1967).
[b] Taking into consideration the age and sex of those who died, and the year of death. The Table shows no evidence of excess mortality in the selenium workers from any of these main groups of causes.

cough. A week later he had a recurrence of the fever with chest pain and the signs of bronchopneumonia. After 4 more days, the fever disappeared and 9 weeks later he was fit and well again. A man, 5 h after cleaning out apparatus used for making hydrogen selenide, developed severe dyspnoea, diffi- culty in expiration, painful cough, and yellowish sputum (Bonard & Koralink, 1958). His condition became worse and the next day he was admitted to the hospital. He was cyanosed, had severe expiratory dyspnoea, discrete rhinitis and conjunc- tivitis, and was in considerable distress. He had no headache or visual troubles. His pharynx was hyperaemic, his buccal mucosa was dry, and he had a furred tongue. Four days later, the chest roentgenogram was normal, but lung function tests were depressed. Rohmer et al. (1950) reported the case of a chemist exposed to a high level of hydrogen selenide, calling attention to the development of severe hyperglycaemia that could only be controlled by increasingly large doses of insulin. A 24-year-old white man accidentally inhaled hydrogen selenide while transferring the gas from one cylinder to another (Schecter et al., 1980). He immediately experienced burning in his eyes and throat followed by coughing and wheezing. He was given oxygen and improved over a period of 2 h. However, 18 h later, because of recurrent cough and progressive dyspnoea, he was hospitalized. Pneumomediastinum developed in this patient and pulmonary function tests revealed restrictive and obstructive airways disease, which slowly improved.

Glover (1970), who reviewed 8 cases of acute hydrogen selenide poisoning (5 laboratory workers and 3 from industrial accidents; some of them probably identical with those mentioned above), and summarized the following sequence of events. The first effects observed after exposure are signs of irritation of the mucous membranes, i.e., running nose, running eyes, cough, sneeze, followed by a slight tightness of the chest. This clears, and there may be a latent period of 6 - 8 h. The patient, who has often returned home by this time, usually wakes up in bed breathless with signs and symptoms of pulmonary oedema. Glover, at the same time, underlined the importance of oxygen therapy in these intoxi- cations, particularly with regard to the prevention of the development of pulmonary oedema.

Lazarev & Gadaskina (1977) referred to 5 cases of hydrogen selenide intoxication at an exposure level of approximately 7 mg/m^3. These cases manifested nausea, vomiting, vertigo, and extreme tiredness, in addition to effects on the respira- tory tract and conjunctiva. The same author mentioned one case of laboratory exposure to hydrogen selenide in which a chemist developed lung oedema and long-lasting cyanosis with respiratory difficulties. On the 22nd day, thrombophlebitis

was observed and within 52 days signs of myocardial damage
were noted.

(b) Prolonged and/or repeated exposure

An important point when considering prolonged low-level
exposure to hydrogen selenide was recognized by Dudley &
Miller (1941). In their experience, acute exposure levels of
5 µg/litre resulted in such eye and nose irritation in
workers that the men could not continue their duties without
the protection of a gas mask. At 1 µg/litre, this irritation did not occur for several minutes, though the presence of
the gas was detected by its odour. However, continued exposure at these levels results in olfactory fatigue so that
workers lose their ability to smell the gas and are thus
disarmed against subsequent increases in levels of the gas in
the work-place.

The effects of prolonged exposure to hydrogen selenide
were described by Buchan (1947) in workers who were engaged in
a process involving the etching of a pattern on steel stripping. In this operation, steel stripping is passed between
rubber imprinting wheels with a raised pattern on the
periphery. The pattern is in constant contact with a wick
immersed in the etching ink. After imprinting, the steel
stripping passes between felt wipers and dips into a rust-
preventive oil bath passing then to a reel. The odour of
hydrogen selenide was detectable at the etching machine and
was particularly offensive in the sludge of the oil bath.
Selenium was identified, qualitatively, in the sludge.
Apparently, the hydrogen selenide was generated as a result of
the reaction of the excess selenium deposited on the steel
stripping in an acid medium, i.e., the oil bath, which was
further acidified by contamination with the etching ink
carried by the steel stripping. Five out of 25 workers
complained of nausea, vomiting, metallic taste in the mouth,
dizziness, extreme lassitude, and fatigability. The symptoms
lasted 2 weeks and were said to be increasing in severity.
Until one month before the onset of symptoms, an etching ink
containing nitric acid and silver was used, but then a new ink
containing selenious acid at 52 mg selenium/litre was substituted.

Air samples were taken at 6 representative sampling
points. During the analysis, qualitative detection of selenium
was made, but, because of the lack of sensitivity of the
titrimetric method used, it was not possible to measure the
selenium quantitatively, and the results were recorded as less
than 0.2 parts per million. Spot and 24-h specimens of
workers' urine contained 0 - 131 µg selenium/litre or an
average of 61.6 µg/litre. So-called control specimens

contained similar amounts, but the controls were the profes-
sional staff collecting the air samples and they had been
exposed to the same environment as the workers for almost
5 h. Also, it should be realized that hydrogen selenide
produces symptoms at exposures too low to increase the urinary
excretion above normal values. When the selenium ink was
replaced by silver ink, there was a gradual regression of
symptoms and, within 6 months, all complaints had ceased and
there was no recurrence.

The Task Group was not aware of any other reports dealing
with the prolonged exposure of human beings to hydrogen
selenide.

8.1.2.1.3 Selenium oxychloride

Less than 5 µg of pure selenium oxychloride on the skin
of the hand resulted in a painful reaction within a few
minutes, exythema surrounding the central necrotic area
(Dudley, 1938). Within 1 month, the skin defect healed with a
scar.

8.2 Low Selenium Intake

8.2.1 Evidence supporting the possible essentiality of selenium in man

Beneficial nutritional effects of selenium have been
observed in several different species (Schwarz, 1976), and
specific signs of selenium deficiency in the presence of
adequate intake of vitamin E have been demonstrated in rats
and chicks (section 7.2.1). Thus, the question of whether
selenium is essential for man arises. The identification of
any human disease due to low selenium intake is difficult,
because selenium deficiency in animals is characterized by a
wide variety of signs involving several different organ
systems.

Although a clear-cut pathological condition attributable
to selenium deficiency alone has not yet been demonstrated in
human beings, certain evidence suggests that selenium may be
essential for man.

For example, purified glutathione peroxidase, from human
red blood cells, contains quantities of selenium similar to
those found in the enzyme isolated from animals (Awasthi et
al., 1975). Moreover, selenium is necessary for the optimal
growth of human fibroblasts in purified cell culture media
(McKeehan et al., 1976). Blood-selenium levels are depressed
in children suffering from kwashiorkor (Burk et al., 1967;
Levine & Olson, 1970) and Hopkins & Majaj (1967) obtained a
reticulocyte response in malnourished infants treated with a

physiological dose of selenium (25 µg selenium, as sodium selenite). Finally, on the basis of the distribution of selenium in various human tissues, Liebscher & Smith (1968) concluded that selenium must be an essential element for man.

8.2.2   Signs and symptoms of low intake

The greatest possibility of a hazard due to inadequate selenium intake would be expected in low-selenium areas. Nutritionists and public health officials in several countries are aware of the low-selenium status of their populations and are attempting to identify any human health problems associated with it. For instance, in the south island of New Zealand, one of the first regions where a low selenium status in animals was recognized, some farmers, noting supposed similarities between their own symptoms and those of the white muscle disease affecting their livestock, claimed improvement in their muscular complaints after self-medication with selenium (Hickey, 1968). However, such anecdotal reports are not supported by the results of 3 separate trials involving a total of 120 patients suffering from "muscular complaints", carried out in several low-selenium areas of the south island of New Zealand (Robinson et al., 1981). In these studies, the patients were given either sodium selenite or selenomethionine at a number of dose levels and schedules for various periods of time. Some subjects also received vitamin E with the selenium supplement. In all subjects who received selenium, blood-selenium levels and glutathione peroxidase activity increased, whereas little change was seen in the control group. Clinical assessment of muscular symptoms showed that approximately equal numbers of patients in the test and control groups exhibited an improvement in their muscular condition. On the basis of these results, the authors concluded that there was no evidence of any response, under the conditions of the trials, to selenium supplements for the relief of muscular complaints.

Total parenteral nutrition (TPN) fluids for intravenous feeding contain very low levels of selenium (section 4.1.1.1), and patients sustained by such techniques would seem to be at risk of developing selenium deficiency. One such patient on TPN in New Zealand developed muscular discomfort that disappeared after selenium supplementation (van Rij et al., 1979). This patient was a 37-year-old female who lived in a rural area of the south island of New Zealand where the soils were low in selenium and there was a history of endemic white muscle disease in sheep. She presented with a perforated small intestine with peritonitis, after radiotherapy 2 years previously for carcinoma of the cervix. Five days after abdominal exploration and the start of intensive antibiotic therapy she developed enterocutaneous and vaginal fistulae. Gastric stress ulcer followed and necessitated a 1.5-litre blood transfusion. Elevated temperatures persisted with intraabdominal sepsis. Ten days after admission to hospital, TPN was begun and resulted in a general improvement and a 6 kg weight gain over the next 20 days. Regular plasma and albumin infusions treated hypoproteinemia. After 20 days of TPN, early clinical signs of fatty acid deficiency (dry flaky skin on hands and feet) were noted, which responded rapidly to intralipid infusion. After 30 days of TPN, the patient complained of increasing bilateral muscular discomfort in her quadriceps and hamstring muscles. Muscle pain was present at rest as a persistent ache and with tenderness on palpation. The muscle pain was aggravated by walking until she found it distressing, even to move beyond her room. On examination, there was tenderness of the quadriceps, hamstrings, and less markedly of the calf muscles of both legs. Both active and passive movements of these muscle groups were painful, partic- ularly of the hamstrings. The upper limb girdle was unaffected. A generalized muscle wasting of all the limbs was observed after the prolonged catabolic stress, despite TPN. No muscle fasciculation or neurological deficits were observed.

Supplementation with selenium was begun with no other modification to the patient's management. Each day 100 µg of selenium, as selenomethionine, was infused intravenously with the TPN solution. During the next week, muscle pain at rest, tenderness to palpation, and pain on active and passive movement disappeared. A return to full mobility followed. This symptomatic response associated with selenium supplement- ation plus the extremely low blood-selenium levels initially observed in this patient (discussed further in section 8.2.4) led the authors to conclude that this case could be the first clinical report supporting the essential role of selenium in human nutrition.

Because of their increased metabolic requirements and faster growth rates, infants and children might be partic- ularly vulnerable to selenium deficiency. McKenzie et al. (1978) analysed blood from 230 healthy adults and 83 healthy children from various areas of New Zealand and found that children in Auckland and Tapanui had lower blood-selenium concentrations (0.064 and 0.048 mg/litre) than adults from the same areas (0.083 and 0.060 mg/litre). Red-blood-cell glutathione peroxidase activity was also lower in Auckland children than in adults (10.6 versus 12.9 units/g haemo- globin). A specific population of infants and children that might be especially at risk regarding low selenium intake includes those who suffer from phenylketonuria (PKU) and maple syrup urine disease (MSUD) and consume only special synthetic diets that are very low in selenium (section 4). McKenzie et al. (1978) reported that the mean blood-selenium level in 12 such children was 0.038 ± 0.013 mg/litre. One 13-year-old patient had 0.016 mg selenium/litre whole blood and 0.009 mg selenium/litre plasma, but clinical examination indicated that he was in good health. Lombeck et al. (1978) found that the serum-selenium content in 36 children receiving diet therapy for PKU and MSUD in the Federal Republic of Germany ranged from 0.007 - 0.028 mg/litre. The selenium content of the hair was lower in the patients (0.062 mg/kg) than in healthy children (0.429 mg/kg) and the erythrocyte glutathione peroxidase activity was reduced in comparison with normal values (4.6 versus 8.8 units/g haemoglobin). And yet all the patients thrived during the time of observation and did not exhibit any increased rate of haemolysis or oxidation of haemoglobin to methaemoglobin after incubation of their erythrocytes with sodium azide.

Gross (1976) studied 4 groups of premature infants who were fed 4 different formulae based on cow's milk containing a high concentration of polyunsaturated fatty acid (PUFA), with and without iron, or a low PUFA concentration, with and without iron. The tocopherol content was the same in all 4 formulae and was judged adequate in terms of maintaining serum-vitamin E levels. Both glutathione peroxidase activity and plasma-selenium levels were similar in all 4 groups. The former declined from 4.2 units/g haemoglobin at one week of age to 2.7 units/g haemoglobin at 7 weeks of age, whereas the latter declined from 0.08 to 0.035 mg/litre. Although all infants exhibited the anaemia typical of the prematurely newborn, the decreases in haemoglobin and increases in reticulocyte levels were greatest in the group of infants given the formula high in PUFA and iron. These haemolytic events in vitamin E-sufficient premature infants fed a diet rich in PUFA and iron were thought to be due to the oxidative stress of the formula coupled with the poor nutritional status of the infants with regard to selenium (Gross, 1976).

8.2.3   Dietary levels consistent with good nutrition

8.2.3.1   Quantitative estimates

Since no clear-cut pathological condition attributable to selenium deficiency alone has yet been observed in man, it is not possible to define a precise dietary requirement level for human beings. However, 0.1 - 0.2 mg selenium/kg diet is a nutritionally generous level for most species of animals (US NAS/NRC, 1971). If these animal data are extrapolated, a 70-kg man consuming 500 g of diet per day (dry basis), would need a daily intake of 50 - 100 µg. The US National Research Council has estimated that the safe and adequate range of the daily intake of selenium for adults is 50 - 200 µg, with correspondingly lower intakes for infants and children (Table 53). On this basis, the recommended intake for a 70-kg man would be equivalent to 0.7 - 2.8 µg/kg body weight per day. Any daily intake within the recommended range is considered adequate and safe, but the recommendations do not imply that intakes at the upper limit of the range are more desirable or beneficial than those at the lower limit. The lower recommended intake for infants and children is consistent with the observation that supplementation of malnourished children with sodium selenite at 30 µg selenium daily, produced weight gain and reticulocyte responses without any untoward signs (Hopkins & Majaj, 1967).

Table 53. Estimated safe and adequate range of selenium intake[a]

Group	Age (years)	Daily selenium intake (µg)
Infants	0 - 0.5	10 - 40
	0.5 - 1	20 - 60
Children	1 - 3	20 - 80
	4 - 6	30 - 120
	7+	50 - 200
Adults		50 - 200

[a] Adapted from: US NAS/NRC (1980).

Three approaches have now been used to estimate human nutritional requirements for selenium. Nutritionists have long used metabolic balance studies to determine the human requirements for a variety of minerals. In the case of selenium, healthy North American men needed about 80 µg dietary selenium/day to maintain balance, but women needed only 57 µg/day (Levander & Morris, 1984). The difference between men and women in the selenium intake needed to achieve balance was considered to be due to differences in body weight. Expressing the balance data on a body weight basis revealed that both men and women needed about 1 µg selenium per kg body weight per day to stay in balance. However, other research groups showed that selenium balance in New Zealand women or Chinese men could be reached on intakes as low as 27 and 9 µg/day, respectively (Stewart et al., 1978; Luo et al., 1985). This demonstrates the great effect of prior dietary-selenium intake on the amount of selenium needed to

achieve balance in people and shows that balance studies may
not be valid techniques for estimating human selenium
requirements.

Yang et al. (1985) estimated human selenium requirements
on the basis of a comparison of dietary intakes in areas with
and without Keshan disease. Dietary-selenium intakes of 7.7
and 6.6 µg/day in endemic and 19.1 and 13.3 µg/day in
non-endemic Keshan disease areas were reported for adult men
and women, respectively. These estimates should be considered
minimum daily adult requirements for selenium.

The depletion/repletion study is another approach used by
nutritionists to estimate human selenium requirements. The
relatively large body pool of selenium in North Americans
prevented their plasma-selenium levels from dropping to values
commonly found in persons from low-selenium areas (Finland,
New Zealand), even when depleted for almost 7 weeks (Levander
et al., 1981a,b). Chinese men of naturally-low-selenium
status (dietary intake about 10 µg/day) were given graded
supplements of selenomethionine and their plasma-glutathione
peroxidase activity was followed (Yang et al., 1985). The
activity of the enzyme plateaued at the same level in all men
receiving 30 µg or more of supplementary selenium daily.
From this study, a physiological selenium requirement of about
40 µg/day (diet plus supplement) was suggested for Chinese
adult males, which may require adjustment to account for body
weight differences in other populations including women.

8.2.3.2 Nutritional bioavailability

None of the studies discussed in section 8.2.3.1 have
specifically addressed the question concerning the nutritional
bioavailability of the selenium in foods for human beings.
Using an animal model, Douglass et al. (1981) found that the
selenium in freeze-dried, water-packed canned tuna for human
consumption was only 57% as effective as selenite in restoring
liver glutathione peroxidase activity in rats previously
deficient in selenium, whereas selenium as cooked freeze-dried
beef kidney or seleniferous wheat had 97 and 83% of the
activity of selenite, respectively. The selenium in the tuna
was also less effective than selenite in raising hepatic-
selenium levels in the deficient rats. Similar results
concerning the relative bioavailability of the selenium in
tuna compared with wheat were obtained by Alexander et al.
(1983) in rats, using the slope-ratio technique. On the other
hand, Chansler et al. (1983) found that the selenium in
mushrooms was only about 4% as available as that in selenite
for restoring hepatic glutathione peroxidase activity in
selenium-depleted rats. It has been pointed out that, in
addition to absorption and retention, factors such as

utilizability within the body are apparently important in determining selenium bioavailability (Levander, 1983).

Although data are accumulating on the absorption by human beings of the selenium in various compounds or in foods (section 6.1.1.2), there have been few studies examining the bioavailability (i.e., utilization, consisting of transport, conversion to a metabolically-active form, retention, etc., in addition to absorption) of the selenium in foods for human beings. One such study was carried out in a low-selenium area of central Finland (Levander et al., 1983). Three groups of 10 men of low-selenium status (mean plasma-selenium level of 70 µg/litre) were supplemented with 200 µg of selenium, daily, as selenium-rich wheat, selenium-rich yeast, or sodium selenate, for 11 weeks. Twenty unsupplemented subjects served as controls. Plasma-selenium levels increased steadily in the wheat and yeast groups for 11 weeks to around 160 µg/litre with no sign of plateauing, whereas, in the selenate group, plasma-selenium plateaued at about 110 µg/litre, after 4 weeks. Red blood cell-selenium levels also increased steadily in the wheat and yeast groups for 11 weeks from 90 to 190 µg/litre, again with no sign of plateauing. Red blood cell-selenium levels were unaffected in the selenate group. Platelet glutathione peroxidase activity (glutathione peroxidase is an index of selenium status) (section 7.2.4.4) roughly doubled after 4 weeks of supplementation with wheat or selenate and then plateaued. Platelet glutathione peroxidase increased more slowly in the yeast group. Plasma-glutathione peroxidase activity did not respond to selenium supplementation. Ten weeks after the supplements were stopped, platelet glutathione peroxidase remained higher in the wheat and yeast groups than in the selenate group. This suggested that the selenium in yeast or wheat was, to some extent, deposited in the tissues in a form that could be used later for glutathione peroxidase biosynthesis, once the dietary supplement was discontinued. The results of this bioavailability trial indicated that there are several different aspects to the nutritional availability of selenium. Complete assessment may require several measurements including: short-term platelet glutathione peroxidase activity, to estimate immediate availability; medium-term plasma-selenium levels, to determine retention; and long-term platelet glutathione peroxidase activity, after discontinuation of supplements, to estimate the convertibility of tissue-selenium stores to metabolically active selenium.

Griffiths & Thomson (1974) also noted that the blood-selenium levels of adults from the USA declined rapidly on arrival in New Zealand, but, after a year, their levels were still higher than the mean value for permanent New Zealand residents (Fig. 1). When selenium levels were measured in the

whole blood, erythrocytes, and plasma of postoperative
surgical patients receiving TPN in New Zealand, selenium
concentrations decreased as TPN was continued (Table 54) (van
Rij et al., 1979). The plasma-selenium concentration of the
New Zealand TPN patient, discussed in section 8.2.2, was 0.025
mg/litre and fell to 0.009 mg/litre just before selenium
supplementation was begun. The lowest value for blood-
selenium concentration in areas of China not affected by
Keshan disease was around 0.040 mg/litre, while, in affected
areas, the blood-selenium concentration often dropped below
0.010 mg/litre (Keshan Disease Research Group, 1979b).

Table 54. Effects of total parenteral nutrition on the concentration of
selenium in whole blood, erythrocytes, and plasma of postoperative
surgical patients in New Zealand[a]

Duration of TPN (days)	Selenium concentration		
	Whole blood (mg/litre)	Erythrocyte (mg/litre)	Plasma (mg/litre)
0	0.050 ± 0.006	0.077 ± 0.008	0.031 ± 0.004
10–20	0.040 ± 0.003	0.060 ± 0.004	0.022 ± 0.002
> 20	0.025 ± 0.003	0.043 ± 0.003	0.015 ± 0.004

[a] Adapted from: van Rij et al. (1979).

The urinary excretion of selenium by New Zealand residents
is quite low, reflects their low intake, and is related to
whole blood-selenium levels (Fig. 11). Similar relationships
between 24-h urinary-selenium excretion and plasma-selenium
concentrations have been observed in New Zealand residents,
New Zealand patients with TPN, and Swedish patients with TPN
(van Rij et al., 1979).

8.2.4   Blood and urine levels typical of low intake

In separate but simultaneous publications, Griffiths &
Thomson (1974) and Watkinson (1974) reported that the mean
selenium content of whole blood from New Zealand subjects was
0.068 and 0.069 mg/litre, respectively. However, even in New
Zealand, where the residents generally have a low selenium
status, it is apparently possible to observe regional
differences in blood-selenium levels (Table 55).

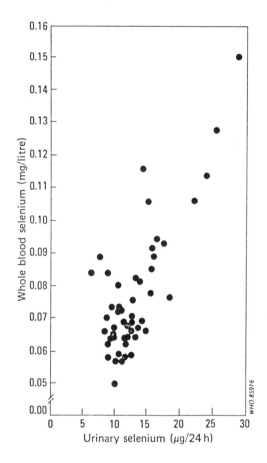

Fig. 11. Correlation of total 24-h urinary excretions of selenium with whole blood-selenium concentrations in New Zealand subjects. From: Griffiths & Thomson (1974).

8.2.5   Relationship between blood-selenium levels and erythrocyte-glutathione peroxidase activity

There is an excellent correlation between human whole blood-selenium concentrations and glutathione peroxidase activity, at concentrations below about 0.10 mg/litre (Thomson et al., 1977b). However, above this concentration, activity of the enzyme is not noticeably increased (Fig. 12), which suggests either that this concentration of selenium is optimal

Table 55. Selenium concentration in whole blood of human beings residing in different areas of New Zealand[a]

Area of New Zealand	Blood-selenium concentration (mg/litre)
Heriot	0.057 ± 0.012
Dunedin	0.062 ± 0.013
Kurow	0.070 ± 0.009
Oamaru	0.074 ± 0.012

[a] Adapted from: Griffiths & Thomson (1974).

and that an intake that maintains this concentration is adequate for function as measured by glutathione peroxidase activity, or, that above this concentration, other factors might play a greater role in influencing glutathione peroxidase activity. Rea et al. (1979) showed that there was also an excellent correlation between human erythrocyte-selenium concentrations and whole blood-glutathione peroxidase activity, as long as the former was less than 0.14 mg/litre (Fig. 13). Schrauzer & White (1978) did not observe any correlation between glutathione peroxidase activity and selenium concentrations in the blood of subjects whose blood-selenium levels were all over 0.10 mg/litre. Moreover, the activity of the enzyme did not increase after these subjects were supplemented with a selenized yeast preparation, even though blood-selenium levels responded to such treatment. Only about 10% of the total selenium in human red cells is associated with glutathione peroxidase (Behne & Wolters, 1979) whereas, in sheep red cells, most of the selenium appears to be associated with the enzyme (Oh et al., 1976b). Thus, the role of selenium in glutathione peroxidase may not be the only function of the element. Whether the non-glutathione peroxidase selenium in human red blood cells truly represents other functional forms of the element or is merely non-functional selenium non-specifically incorporated into tissue proteins cannot be answered at this time. But the usefulness of the glutathione peroxidase assay as a means of assessing selenium intake in human beings whose whole blood-selenium concentration exceeds 0.10 mg/litre currently appears to be an open question.

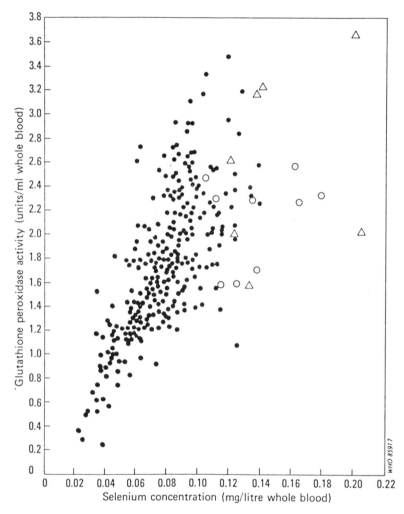

Fig. 12. Relationship between whole blood-selenium concentration and whole blood-glutathione peroxidase activity in New Zealand residents. From: Thomson et al. (1977b).

(○) 264 New Zealand residents.
(⊙) 9 New Zealand residents returned from overseas visits.
(△) 7 new settlers to New Zealand.

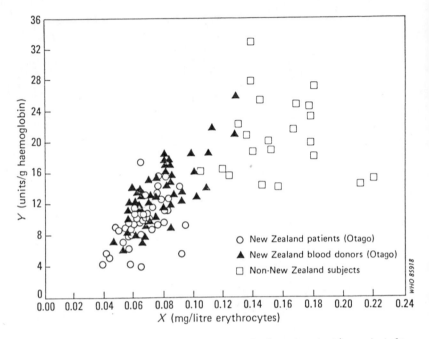

Fig. 13. Relationship between erythrocyte-selenium concentration and whole blood-glutathione peroxidase activity. From: Rea et al. (1979).

8.2.6 Attempts to associate low selenium intake with human diseases

8.2.6.1 Keshan Disease

Results of research in China have suggested a relationship between low selenium status and the prevalence of Keshan disease, an endemic cardiomyopathy that primarily affects children (Keshan Disease Research Group, 1979a,b). Cases of Keshan disease were recorded as early as 1907 in Heilongjiang Province of northeastern China (Gu, 1983). Since the etiology of the disease was not known, it was named after the locality in which it was originally observed, Keshan County.

Epidemiologically, the disease exhibits a regional distribution and occurs in a belt-like zone reaching from northeastern China to the southwestern part of the country. There is a marked seasonal fluctuation in the disease with more cases appearing during winter in the north and during summer in the south. There is also a great annual variation in the incidence of Keshan disease. Recently, the incidence has decreased sharply such that from 1978 to 1980 less than one death was reported per 100 000 members of the population (Gu, 1983). The overall recent decline in the incidence of Keshan disease has been attributed, at least in part, to the general increase in the living standards of the people, such as better sanitation, more medical attention, and improved quality of the diet (He, 1979). There is also a shifting of epidemic foci from year to year. Rural peasants constitute the population at risk with children below 10 years of age and women of child-bearing age most susceptible. According to Gu (1983), migrants from non-affected areas will not contract the disease unless they have lived in the endemic area for at least 3 months.

The criteria for diagnosing Keshan disease include acute or chronic cardiac insufficiency, heart enlargement, gallop rhythm, arrhythmia, and ECG changes (Chen et al., 1980). The disease is classified into four types: acute (cardiogenic shock), subacute, chronic (low output pump failure), and latent (normal heart function but mild enlargement). There is no symptom or sign specific for identifying the disease (Gu, 1983). Histopathologically, Keshan disease is characterized by multifocal necrosis and fibrous replacement of the myocardium.

In the past, many different hypotheses were advanced in an attempt to explain the etiology of Keshan disease. For example, there were theories of poisoning due to rhodamine, silica, digitalis, carbon monoxide, barium, nitrite, and Fusarium mycotoxins. However, clear-cut proof of these toxicants as a cause of the disease has not been obtained despite much analytical effort (Yang et al., 1984). Several nutritional problems have also been proposed to play a role in Keshan disease such as deficiency of protein, lipids, thiamin, magnesium, or molybdenum. A hypothesis concerning Keshan disease etiology was proposed linking the disease to selenium deficiency, after it was noted that severely endemic areas coincided with areas where the incidence of enzootic selenium deficiency diseases in farm animals was also high (Zhu et al., 1981). On this basis, selenium supplements were used as a preventive measure for Keshan disease. This hypothesis is very appealing since cardiomyopathy is a prime feature of selenium-vitamin E deficiency in many species of animals, including cattle, sheep, and swine.

Since the initial proposal of the selenium-Keshan disease connection, much evidence has been gathered in support of this concept. For example, the average blood-selenium content was 0.021 ± 0.001 mg/litre for the affected areas and 0.095 ± 0.088 mg/litre for non-affected areas (Yang et al., 1984).

The average hair-selenium content was below 0.12 mg/kg in affected areas, whereas hair-selenium levels in neighbouring but unaffected areas ranged between 0.12 and 0.2 mg/kg. Average hair-selenium contents in areas removed from the affected belt were between 0.25 and 0.6 mg/kg. The selenium level in several staple foods (rice, maize, wheat, soybeans, black beans, and sweet potatoes) was lower in areas affected with Keshan disease than in unaffected areas. It was stated that an area could be considered to be unaffected wherever the selenium content of grains was 0.04 mg/kg or more. The Chinese workers stated that the amount of selenium needed to prevent the disease was about 20 μg/day (Yang et al., 1984).

In some affected areas, there were highly localized pockets (so-called "safety islands") that were free from the disease. Apparently, these islands were protected because of the higher selenium content of their crops in the immediate vicinity. For example, the average selenium contents of rice and soybeans in one such island were 0.020 and 0.025 mg/kg, respectively, whereas the values for the corresponding crops in a nearby affected area were 0.0078 and 0.0057 mg/kg. It was also noted that the children in the unaffected spot liked to catch shrimp from local streams and that the selenium content of these dried shrimp was as high as 1 mg/kg (Keshan Disease Research Group, 1979b).

These relationships between selenium and Keshan disease led the Chinese workers to conduct a randomized intervention trial to test the possible prophylactic effect of selenium against this condition in the population at risk (i.e., children 1 - 9 years old). In a trial in 1974, 4510 children took sodium selenite and 3985 children took the placebo (Table 56). The treated children took 0.5 mg sodium selenite per week if 1 - 5 years old, or 1.0 mg per week if 6 - 9 years old. The morbidity rate due to Keshan disease was 1.35% in the placebo group (54 cases out of 3985 children) but only 0.22% in the treated group (10/4510). Since a significant difference was also shown in the 1975 trial (0.95% morbidity rate in the placebo group compared with 0.1% in those treated) the placebo groups were abolished in 1976 and 1977. As a result, the case rate dropped to 0.034% and 0% in these 2 years, respectively. However, in 1976, there was one case out of 212 children who failed to take the treatment.

No untoward side effects due to the sodium selenite were observed, except for some individual cases of nausea, which could be overcome by taking the medicine after meals. Physical examinations and liver function tests indicated that the liver was undamaged after continuous ingestion of the selenium tablets for 3 - 4 years.

Since 1976, more extensive intervention trials with sodium selenite have been carried out in 5 counties in the same area

Table 56. Effect of selenium on Keshan disease in children[a]

Treatment	Year	Number of subjects	Number of cases	Outcome of alive cases			Death
-----------	------	--------------------	-----------------	Turned latent	Improved	Turned chronic	
Placebo	1974	3985	54	16	9	2	27
	1975	5445	52	13	10	3	26
Sodium	1974	4510	10	9	0	1	0
selenite	1975	6767	7	6	0	0	1

[a] Adapted from: Keshan Disease Research Group (1979a).

of Sichuan Province (Yang et al., 1984). All children, 1 to 12 years of age, in some of the most severely affected communes, were treated with selenium as described above, while untreated children in nearby communes served as controls. The incidence rate of Keshan disease in the selenium-treated children was lower in each year of the five-year period than among the untreated children (Table 57).

Table 57. Keshan disease incidence rates in selenium-treated and untreated children in five countries of Sichman province[a]

Year	Treated children			Untreated children		
------	Number of subjects	Number of cases	Incidence (per 1000)	Number of subjects	Number of cases	Incidence (per 1000)
1976	45 515	8	0.17	243 649	448	2.00
1977	67 754	15	0.22	222 944	350	1.57
1978	65 953	10	0.15	220 599	373	1.69
1979	69 910	33	0.47	223 280	300	1.34
1980	74 740	22	0.29	197 096	202	1.07
Total	323 872	88	0.27	1 107 568	1713	1.55

[a] Taken from: Yang et al. (1984).

Selenium intervention has proved to be very effective in the prophylaxis of Keshan disease and it is very likely that selenium insufficiency plays an important role in its etiology (Yang et al., 1984). Nevertheless, the Chinese workers recognized that there were certain epidemiological character-istics of the disease, which suggested that additional etiological factors were involved. The selenium hypothesis, for example, does not adequately explain the seasonal variation in the disease, the occurrence of epidemic years, or the annual shifting of epidemic foci. Such characteristics are more compatible with an infectious theory and, in fact, a hypothesis that the disease is a form of viral myocarditis has been put forward (He, 1979). Perhaps the best way to account for all the characteristics of the disease is to assume that the disease has a multifactorial etiology and that a combination of several factors may be involved (Yang et al., 1984). There are some animal studies that favour such a possibility, since selenium-deficient mice were less resistant to the cardiotoxic effects of a Coxsackie B_4 virus isolated from a patient with Keshan disease (Bai et al., 1980). If selenium-deficient human beings are also less resistant to viral infection, this phenomenon could provide a reasonable explanation for many of the apparently conflicting features of Keshan disease. In this view, selenium deficiency would be the fundamental underlying condition that would predispose persons to viral attack, possibly by impairing normal immune function (Yang et al., 1984).

8.2.6.2 Kashin-Beck disease

Kashin-Beck disease is an endemic osteoarthropathy that occurs in eastern Siberia and in certain parts of China, which is characterized as a chronic, disabling, and degenerative osteoarthrosis that mainly involves children (Sokoloff, 1985).

Although the etiology of this disease has not been fully established, present works show that selenium deficiency might be one of the main causes. This concept is based on the following evidence. First, in China most of the endemic areas are located in the same low-selenium zone, from northeast to southwest, as the Keshan disease (section 8.2.6.1) (Tan et al., in press). For the same reason, residents in these areas have low-selenium status characterized by low blood- and hair-selenium levels, low blood glutathione peroxidase activity, and low urinary-selenium excretion. A survey carried out in Heilongjiang province of China (one of the most heavily affected province) showed that the average hair-selenium levels in 151 children in endemic areas (0.096 ± SD 0.026 mg/kg) was significantly lower than that of the 235 children in the non-endemic areas (0.223 ± 0.083 mg/kg) (Wang et al., 1985). The authors concluded that the low selenium status of children was due to the low selenium content of the locally produced staple food. The average selenium content of corn, wheat, and millet in the endemic and non-endemic areas shown in the paper were: 0.0056 ± SD 0.0038 (n = 262) versus 0.015 ± 0.015 (176), 0.0091 ± 0.0096 (225) versus 0.0235 ± 0.022 (120), and 0.0064 ± 0.0053 (14) versus 0.0197 ± 0.0144 (11) mg/kg, respectively. Second, sodium selenite is reported to have both therapeutic and prophylactic effects on this disease. Liang (1985) reported that 325 cases of Kashin-Beck disease in Shaanxi province of China were randomly divided into a treated and control group. The treated group was given sodium selenite (1 mg/week for children 3 - 10 years of age and 2 mg/week for children of 11 - 13 years of age) and the control group was given a placebo. After one year, X-ray examination of the metaphyseal changes of fingers showed that 81.9% of the cases in the treated group had improved, none of the cases were getting worse, and that 18.1% showed no change, while, in the control group, only 39.6% of the cases had improved, 30% were getting worse, and 41.5% showed no change. Li et al. (in press) reported that children of 1 - 5 and 6 - 10 years of age in an endemic area of Gansu province in China were supplemented with 0.5 and 1.0 mg sodium selenite, respectively, per week over a period of 6 years. X-ray examination showed that the incidence of Kashin-Beck disease declined from 42% to 4% after the selenium intervention.

However, other information suggests that other factors in the environment may also play some role in the disease. For example, the possible role of mycotoxin contamination of cereals by certain Fusarium strains in China and high phosphate and manganese contents in the soil, food, and drinking-water in endemic areas of the USSR have been suggested. More studies are required to clarify the relationship between these hypotheses and Kashin-Beck disease.

8.2.6.3 Cancer

(a) Ecological studies

Shamberger & Frost (1969) first pointed out the inverse relationship between selenium levels in forage crops and human blood, and cancer death rates in various regions of the USA. Subsequent papers expanded this concept (Shamberger & Willis, 1971; Shamberger et al., 1976). From a number of comparisons, the cancers frequently found at one time or another to have high mortality associated with low-selenium areas or low blood-selenium levels, included cancer of the tongue, oesophagus, stomach, colon, rectum, liver, pancreas, larynx, lungs, kidneys, bladder, and Hodgkin's disease and lymphoma. However, the high mortality of some of these same cancers and others, was also found to be associated with high-selenium areas or high blood-selenium levels; these include cancer of the lung, prostate, pancreas, breast, lip, skin, eye and dermal melanoma, and leukaemia/aleukaemia. Some of these studies have been criticized as lacking strength and consistency, particularly because the states for which cancer mortality was calculated did not coincide directly with the natural geographical units on which the estimates of selenium levels in forage crops were made, thus leading frequently to misclassification of the different selenium areas (Allaway, 1972, 1978). However, in a study attempting to minimize this problem by using county data, similar inverse correlations were observed between counties classified as intermediate or high for selenium levels in forage, and cancers of the lung, colon, rectum, bladder, oesophagus, pancreas, and all sites combined, for both males and females, and cancers of the breast, ovary, and cervix (Clark, 1985).

Schrauzer (1976) found that the mortality rates due to several cancers, including those of the large intestine, rectum, and breast (so-called Type A cancers) were directly correlated with the consumption of meat, eggs, milk, fat, and/or sugar and were inversely correlated with the consumption of cereals and fish. Just the opposite correlations were found for certain other cancers such as hepatic and stomach cancer. Since cereals and seafoods are good sources of dietary selenium, it was suggested that selenium might be the factor in these foods that protected against Type A cancers. Schrauzer et al. (1977) extended these studies to data from 27 countries including the USA; New Zealand was intentionally excluded. Dietary intake of selenium was found to be inversely correlated with total age-adjusted cancer mortality, r = -0.46 ($\underline{P}$ < 0.01) for males and r = -0.60 ($\underline{P}$ < 0.001) for females. Significant inverse correlations were observed between dietary-selenium intake and mortality from cancers of the colon, rectum, prostate, female breast, ovary, lung (males), and from leukaemia. Weak inverse relationships were found with mortality from cancers of the pancreas ($\underline{P}$ = 0.06), bladder ($\underline{P}$ = 0.1), and skin ($\underline{P}$ = 0.1, males). Other cancers including those of the stomach, oesophagus, and liver did not show any significant direct or inverse correlations with dietary-selenium intake. In this study, dietary-selenium intake was calculated, assuming that the same average concentration of selenium was present in the foods consumed in all countries (the Task Group questioned the validity of such an assumption). Mortality from cancers of the colon, rectum, prostate, lung, skin, bladder (all Type A), and leukaemia showed significant inverse correlations with blood-selenium levels in males (USA excluded). However in the data from the USA, this relationship was not observed for cancer of the prostate, lung, skin, or bladder, and leukaemia, and similar inconsistencies were observed for other cancers and in data for females.

Jansson et al. (1978) postulated an inverse relationship between dietary-selenium and the rate of colorectal and breast cancer, but found a direct correlation between the concentration of selenium in the drinking-water and the rate of colorectal cancer. Moreover, these workers commented that the same statistical associations that indicated a protective effect of dietary selenium against colon, rectum, and breast cancers also indicated an increased risk of liver and stomach cancer due to selenium.

In a recent study in China, Yu et al. (1985) obtained a mean serum-selenium level in the whole blood of 1458 donors from 24 regions, of 107 µg/litre (range 22 - 314 µg/litre). Blood-selenium levels were inversely correlated with age-adjusted total cancer mortality for both males and females, r = -0.64 (P < 0.01) and r = -0.60 (P < 0.01), respectively. Analysis by cancer sites revealed significant negative correlations between blood-selenium levels and stomach and oesophageal cancers in both sexes. On reclassifying regions according to low-, moderate-, and high-selenium areas on the basis of blood levels, significantly lower total cancer death rates were observed in regions with high selenium levels and the mortality from cancer of the stomach, oesophagus, and liver was particularly increased in the low-selenium areas. In an area where primary liver cancer was very common, a statistically significant negative correlation between primary liver cancer incidence and selenium levels in grain was observed (r = -0.623 for maize, and -0.631 for barley corn). An inverse correlation between age-adjusted primary liver cancer and blood-selenium levels of residents in the area was also observed. The authors concluded that the results indicated that selenium might play an important role in the etiology of liver cancer, and that though selenium deficiency was not a cause of primary liver cancer, low selenium intake apparently reduced the ability of the body to withstand cancer-causing stress.

It has been pointed out (Levander, 1986) that the age-adjusted mortality rates for breast cancer and colon cancer reported in Finland are considerably lower than those reported in the USA, despite the well-documented lower dietary-selenium intakes in Finland.

(b) Case-control studies

Shamberger et al. (1973b) reported that blood-selenium levels in patients with cancer of the colon, pancreas, stomach, and in Hodgkin's disease and liver metastases were statistically significantly lower than those in normal controls. However, of 29 patients with rectal cancers, 6 had lower selenium levels than controls, and 23 had normal levels. Similarly, normal levels were observed in patients with breast cancer and in patients with other types of carcinoma.

McConnell et al. (1975), compared blood-selenium levels in 110 patients with carcinomas, 36 patients with primary neoplasm of the reticuloendothelial system, 28 hospitalized patients with no malignancy, and 18 non-hospitalized healthy individuals. The mean selenium concentration for the hospitalized non-malignant patients was 1.49 ± 0.06 mg/kg and for healthy controls 1.48 ± 0.07 mg/kg. The mean level of 1.27 ± 0.03 mg/kg for patients with carcinomas was significantly different from that of the healthy control group ($\underline{P}$ = 0.01). The mean serum-selenium level of 1.14 ± 0.08 mg/kg obtained for gastrointestinal cancer was significantly different from that of healthy controls ($\underline{P}$ < 0.005). In about one-third of the 110 cancer patients having the lowest selenium levels, disseminated tumour, recurrences of the primary lesions, the incidence of multiple primaries, and shortened patient survival time, were more frequently observed than in the third of the patients having the highest serum levels ($\underline{P}$ > 0.001). The mean serum-selenium level for the primary malignancies of the reticuloendothelial system was 1.76 ± 0.24 mg/kg, which though higher, was not significantly different from that for the healthy controls.

Broghamer et al. (1976) reported no difference in serum-selenium levels measured as µg/ml between 110 cancer patients and controls. However, patients with the lowest serum-selenium levels had shorter survival, higher incidence of multiple primary malignancies, higher rate of recurrence of the primary lesion, and were more likely to have dissemination of cancer than those with the highest serum-selenium levels. In a study of 59 patients with primary malignant reticulo-endothelial tumours and controls, Broghamer et al. (1978) found no difference in serum-selenium levels measured as µg/ml between the two groups. McConnell et al. (1980) in their study found statistically significant lower levels of serum-selenium in 35 breast cancer patients compared with a control group of women free of the disease. On the other hand, van Rij et al. (1979) and Robinson et al. (1978a) did not find any differences in the blood-selenium levels of surgical patients with and without cancer. Although nutritional status, age, and severity and duration of disease influenced the selenium levels in the patients studied, low selenium levels were not characteristic for the cancer patients and it was suggested that the low-selenium status of cancer patients was more likely a consequence of their illness rather than the cause of the cancer.

Sundstrom et al. (1984a) found that 44 patients with gynaecological cancer had lower serum-selenium concentrations (1.15 ± 0.04 µmol/litre, $\underline{P}$ < 0.05) and serum-glutathione peroxidase activity (404 ± 13 units/litre, $\underline{P}$ < 0.01) than 56 control subjects (1.25 ± 0.03 µmol/litre and 444 ± 8

units/litre, respectively). It was observed that in association with cytotoxic chemotherapy, selenium alone ($\underline{P}$ < 0.05),
vitamin E alone ($\underline{P}$ < 0.05), and the 2 combined ($\underline{P}$ < 0.001)
decreased the plasma concentration of lipid peroxides; the
combination of selenium and vitamin E also increased the
activity of serum GSH-Px ($\underline{P}$ < 0.01). The authors stated
that during placebo treatment, cytotoxic chemotherapy did not
affect plasma-lipid peroxides but decreased ($\underline{P}$ < 0.001) the
activity of GSH-Px. Selenium inhibited this effect. The
authors concluded that this suggested that the anti-oxidative
mechanisms of patients with these types of cancer might be
defective and that treatment with selenium and vitamin E
resulted in changes in biochemical factors related to lipid
peroxidation.

In another study, Sundstrom et al. (1984b) reported that
patients with ovarian cancer had significantly lower serum-
selenium concentrations (mean 0.93 ± 0.04 μmol/litre, $\underline{P}$
< 0.001) than matched controls (mean 1.22 ± 0.03
μmol/litre). Clinical stage IV patients had lower levels of
selenium (0.82 ± 0.07 μmol/litre) than clinical stage I
and II combined (1.00 ± 0.04 μmol/litre). Moreover,
levels tended to decrease with progressive disease and
increase with remission, probably related to nutrition.

Goodwin et al. (1983) studied blood-selenium levels, and
blood and tissue GSH-Px activity in 50 patients with untreated
cancer of the oral cavity and oropharynx. Mean erythrocyte-
selenium and -glutathione peroxidase were significantly
depressed compared with those in age-matched controls. Mean
plasma-selenium, on the other hand, was significantly elevated
in the cancer group. Also, although not significant, mean
erythrocyte-selenium levels tended to be lower in patients who
had never smoked or who had recently given up smoking. No
correlation between dietary selenium as determined by recall
history and plasma- or erythrocyte-selenium levels in the
cancer patients was observed.

Stead et al. (1985) reported significantly lower
serum-selenium concentrations in 20 patients with cystic
fibrosis, two of whom had cancer, than in controls. The two
cancer cases had mean serum-selenium levels of 1.01
μmol/litre and 0.62 μmol/litre, respectively, compared
with 1.41 ± 0.20 μmol/litre in the controls. Serum-
vitamin E levels were also found to be low in patients, but
did not show any correlation with serum-selenium levels.

(c) Case-control studies within prospective studies

As part of the Hypertention Detection Follow-up Programme,
10 940 men and women with diastolic blood pressures of at
least 90 mm Hg were identified and enrolled between 1973-74,

and followed-up for 5 years (Willett et al., 1983). Venous blood samples were collected from all participants at the beginning of the study. A total of 111 new cases of cancer occurred in the group during the period of observation. For each case, 2 controls without cancer were selected who most closely matched the case in age, sex, race, smoking history, month of blood collection, initial blood pressure, hypertensive medication, randomisation assignment, and (in women) parity and menopausal status. Cases and controls were comparable for most of the confounding factors, although serum-cholesterol and albumin were slightly lower among cases than controls ($\underline{P}$ = 0.05 and 0.06, respectively). Serum-selenium levels did not vary with age or sex in controls, but black subjects had lower selenium levels than white. The mean serum-selenium level in cancer cases (0.129 ± 0.002 mg/litre) was significantly lower than that in controls (0.136 ± 0.002 mg/litre). The $\underline{P}$ value was 0.02. The increased risk of cancer in the lowest quintile of baseline selenium was twice that in the highest quintile (confidence limits 1.1 to 3.3). Cases were too few to examine by cancer site, but a consistent trend of lower selenium levels among cases was observed for the following groups of cancers: lung, breast, prostate, lymphoma/leukaemia, gastrointestinal cancer, and others. However, statistically significant differences were observed only for gastrointestinal cancer. Selenium level remained a significant predictor of risk, even when the effects of serum-retinol, vitamin E, and lipid levels, as well as age, sex, and race were taken into account. On examination of the data according to race, sex, and smoking status, separately, significant differences in the mean serum-selenium between cases and controls were observed in blacks but not whites, in males but not females, and in current smokers, but not in past-smokers or in persons who had never smoked. The risk associated with low selenium was greater among those in the lowest tertile of serum-vitamin E, and a similar inverse relationship was observed for serum-retinol. A very strong effect of low selenium (relative risk = 6.2 for lowest versus highest tertiles) was observed for subjects who had both low serum-vitamin E and -retinol levels. The authors concluded that, although their findings supported the overall hypothesis that low selenium intake increases the risk of cancer, they believed that the observed differences between cancer sites should be treated as hypotheses to be tested in other data sets, and that the differences in the effects of selenium according to age, race, sex, and smoking status, needed to be examined further (Willett et al., 1983).

Salonen et al. (1984b) identified a cohort of 8113 men and women in 1972, randomly selected from 2 counties in Finland, who had no history of cancer in the 12 months preceding this date. The blood samples were collected from each subject at the beginning of the study in 1972, and stored at -20° C. The cohort was followed-up for cancer occurrence and death up to the end of 1978, during which time 43 deaths from cancer occurred and an additional 85 persons developed cancer (total cancers = 128). Each cancer case or cancer death was matched for sex, age, number of cigarettes smoked and total serum-cholesterol, to a control selected from the rest of the same population of 8113 persons. In 1983, selenium concentration was estimated in blood collected from each case and control at the time of enrolment in the study in 1972 before the develop-ment of cancer. The mean concentration of selenium was 50.5 µg/litre (SD = 12.5) for all 128 cases, and 54.3 µg/litre (SD = 11.8) for all controls ($\underline{P}$ < 0.012). This difference between cases and control persisted for the following cancer sites: gastrointestinal, respiratory, and haematological cancers, miscellaneous cancers, and secondary cancers. No such difference was observed for skin cancer, skeletal cancers, and urogenital cancer. Using the lowest (35 µg/litre) and third deciles as cut-off points, relative risks were computed for the 2 lowest levels of serum-selenium, with the highest selenium stratum (45 µg/litre) as the reference. The relative risk of cancer associated with a serum-selenium level of less than 35 µg/litre was 3.0 and that associated with a level of 35 - 44 µg/litre was 2.4 (both statistically significant). The authors concluded that their data provided additional support for the hypothesis that selenium deficiency increases the risk of most non-hormone-dependent cancers in middle-aged persons, though they cautioned that the number of cancer cases in their study was insufficient to draw definite conclusions about the effect of selenium for cancers at specific sites.

Salonen et al. (1985) in a further study of a subgroup of 51 patients, who died from cancer, among the same study population as above, each matched to a control for age, sex, and smoking, obtained a mean pre-follow-up serum-selenium concentration in subjects who died from cancer during the study period of 53.7 ± 1.8 µg/litre and that in controls of 60.9 ± 1.8 µg/litre; the difference was statistically significant. A statistically significant difference in the mean serum-selenium concentration between cases and controls was observed in men ($\underline{P}$ = 0.002), but not in women (half of the male pairs were smokers, but none of the 21 female pairs were smokers). Similarly, the difference in the mean selenium concentration was significant in smokers ($\underline{P}$ = 0.013) but not significant in non-smokers. Years of smoking showed an inverse correlation with serum-selenium in the cases (r = -0.30) but not in controls. Analysis accroding to cancer site revealed that a statistically significant difference in serum-selenium levels between cases and controls was recorded only for respiratory cancers, though a non-significant difference also existed for gastrointestinal sites and other cancers. An association between low serum retinol concentration and increased risk of cancer was observed only among smoking men. This sex difference was attributed to smoking as there were no smoking women in the study.

Vitamin E and serum-retinol did not show any association with a specific cancer site, and although vitamin E had only a weak independent effect on the risk of cancer, it showed a strong synergistic relationship with selenium on the risk of fatal cancer. A serum-selenium concentration of 47 µg/litre or less (the lowest tertile) was associated with a relative risk of death from cancer of 5.8 (95% confidence interval 1.2 - 29.0). The authors concluded that, although their findings indicated that dietary-selenium deficiency increased the risk of cancer, owing to their study design, they could not rule out the possibility that the substances measured in the serum were not truly the protective factors but were merely indicators of some other compounds or nutrients that were directly involved in the causal relationship. Furthermore, the authors recognized that there was need to investigate further the modification and the confounding of effects by sex, smoking, and other factors.

8.2.6.4 Heart disease

Using the same ecological approach discussed above for cancer, Shamberger et al. (1975) concluded that the age-specific death rates for a number of heart diseases were significantly lower in the high-selenium regions of the USA than in the low-selenium regions. However, results of a WHO/IAEA research programme showed no difference in tissue-selenium concentrations between patients who died with, or without, myocardial infarction (Masironi & Parr, 1976). Furthermore, Shamberger (1978) found that the kidney-selenium levels of patients with atherosclerosis and hypertension did not differ from those of patients with a variety of other diseases. The blood-selenium levels of patients with acute myocardial infarction were lower than those of healthy adults, but there was no difference in heart- or liver-selenium levels of patients who died from myocardial infarction and those who died from other diseases (Westermarck, 1977).

A recent case-control study from Finland suggested a possible association between the serum-selenium level and the risk of death from acute coronary heart disease as well as the risk of fatal and non-fatal myocardial infarction (Salonen et al., 1982). The case-control pairs were derived from 11 000 persons residing in eastern Finland, an area with a very high incidence of death from cardiovascular disease. The cases were middle-aged persons who had died of coronary heart disease or other cardiovascular disease or suffered a non-fatal myocardial infarction over a 7-year follow-up period. Attempts were made to control for potential confounding factors by using controls matched for 6 major coronary heart disease risk factors: age, sex, serum-cholesterol, diastolic blood pressure, smoking, and history of angina pectoris, but the cases had slightly higher blood pressure than the controls. The mean serum-selenium levels were 51.8 and 55.3 µg/litre for cases and controls, respectively. A serum-selenium level of less than 45 µg/litre was associated with an increased risk of coronary and cardiovascular death and myocardial infarction. Although few necropsies were done, the authors felt that it was unlikely that the excess cardiovascular mortality observed in their subjects was due to Keshan disease. The authors cautioned that the apparent association between low serum-selenium levels and cardiovascular risk might be spurious since serum-selenium might only be, for example, a marker for other dietary factors more directly related to increased coronary heart disease. The authors also emphasized that, even if their results truly reflected a causal relationship between low selenium intake and increased ischaemic heart disease, most such disease is still due to the other well-known risk factors of elevated cholesterol, high blood pressure, and smoking. Moreover, it was pointed out that any association between low serum-selenium levels and ischaemic heart disease is likely to be of significance only for populations in areas where the dietary intake of selenium is very low.

In 3 other studies from Finland little or no association was found between the risk of death from ischemic heart disease and low selenium status (Miettinen et al., 1983; Salonen et al., 1985; Virtamo et al., 1985). However, a critique of these studies (Salonen, 1985) indicated that the first and third were characterized by low statistical power and in the first the mean serum-selenium level was relatively high compared with typical Finnish values (73 µg/litre), probably due to importation of grain high in selenium during the late 1970s (Mutanen & Koivistoinen, 1983). Salonen (1985) stated that all 4 Finnish studies discussed above supported the concept of an increased risk of ischemic heart disease due to low selenium intake as indicated by serum-selenium levels of less than 60 µg/litre.

In the USA, an inverse correlation was reported between the plasma-selenium level and the severity of coronary atherosclerosis as documented arteriographically (Moore et al., 1984). The mean plasma-selenium levels of patients with "zero-vessel" disease (no visible narrowing as much as 50% of any coronary arteriol lumen) or "three-vessel" disease (as much as a more than 50% narrowing in the three major coronary arteries or their branches) were 136 ± 7 or 105 ± 4 µg/litre, respectively. On the other hand, neither Ellis et al. (1984) in the United Kingdom nor Robinson et al. (1983) in New Zealand were able to demonstrate any correlation between the traditional risk factors for cardiovascular disease and blood-selenium levels or glutathione peroxidase activity.

9. EVALUATION OF THE HEALTH RISKS ASSOCIATED WITH EXCESSIVE OR DEFICIENT SELENIUM EXPOSURE

9.1 The Need to Consider the Essentiality of Selenium in the Health Risk Evaluation

The data presented in the preceding sections show evidence that selenium is a functional component of an enzyme (glutathione peroxidase) in animals and man, and can prevent certain diseases in animals and responsive conditions in man. The Task Group concluded that selenium meets the criteria of essentiality for man. As is true for all essential elements, not only deficient but also excessive exposure results in adverse health effects.

The effects of selenium deficiency as well as toxicity are well known in several animal species. In contrast to the effects seen in animals, health effects in man resulting from deficiency or excess of selenium are less well defined, but available evidence has been described. They can occur at low and excessive exposure levels that may be expected to correspond broadly with those having deleterious effects on the health of animals. Between these extremes is a range of safe and adequate exposures (intakes) that can be defined as free from toxicity and adequate to meet nutritional requirements. Exposures outside this range increase the risk of adverse health effects. Therefore, the health risk evaluation of both selenium deficiency and excess is important. However, it should be noted that the safe and adequate range may be modified by certain dietary and other environmental conditions.

The aim of the Task Group was not to evaluate the need for, and safety of, medication by selenium compounds as a preventive or therapeutic measure. However, the Task Group felt that some of the observations presented might be of relevance whenever such a question might be addressed by any other body responsible for the appropriate risk-benefit evaluation of administration of selenium compounds to human beings.

9.2 Pathway of Selenium Exposure for the General Population

Reproducible and accurate methods are available for the collection of environmental and biological samples and the measurement of their selenium content. These methods, though they require special skills and equipment and are time-consuming, have been employed to assess human selenium exposure. However, the data are incomplete and there are no data available on selenium exposure levels for the general population in many countries and regions of the world.

15

In spite of these limitations, it can be concluded that, for the general population, the main source of selenium exposure is food. In nutritional surveys, extreme mean values for the calculated selenium intake from food by adult human beings varied from 11 to 5000 µg/day. However, on the basis of the data available from most areas, the Task Group concluded that dietary-selenium intakes usually fall within the range of 20 - 300 µg/day. Exposure via drinking-water is much less and rarely exceeds a few µg/day. Limited data on selenium levels in air indicate that about 0.2 µg can be inspired daily by individual members of the general population.

9.3 Quantitative Assessment of Human Selenium Exposure

9.3.1 Analytical methods for selenium

Several methods of analysis for selenium are available. Those based on fluorometry, neutron activation analysis, or atomic absorption spectrometry have been the most thoroughly studied and used. While simplified and automated procedures have been developed, all of these methods require skilled analysts to obtain consistently accurate results. The analysis of NBS Standard Reference Materials and the exchange of a variety of samples with another laboratory known to be accomplished in analysis for this element, should be used to verify the adequacy of results, prior to any study of human exposure. The method of choice will depend on the availability of equipment and of the pure chemicals that may be required, as well as on certain other factors. Where a high degree of accuracy is required, comparison of results obtained by two or more different techniques is helpful.

The reliability of the estimate of selenium exposure will depend, not only on the adequacy of the method, but also on the adequacy of sampling, storing, subsampling, and preparing the subsamples for selenium determination. Failure to plan for, and observe, proper practices in these steps can negate the reliability of results by even the most highly accurate method of measurement.

9.3.2 Food intake data

Because of the wide variation in the selenium content of foods in different regions, special techniques must be used to assess human selenium exposure on the basis of food intake data. The most accurate method is to determine selenium levels in duplicate diets (food-on-the-plate-method) made from the same food that is consumed by the subjects. This method may be suitable for studies in small subpopulations and where a high degree of accuracy is necessary; it has the disadvantage of being expensive and time-consuming.

The use of a nutrition survey approach with calculation of selenium intake from food tables can be used if an important rule is observed, i.e., the food tables used must be formulated from data on the selenium contents of the food sources of the population being studied and of the food as eaten by the subjects under study. This is important to avoid the errors introduced by the wide variation in selenium content of a given foodstuff, depending on its origin.

9.3.3   Blood-selenium

The selenium content of whole blood, serum, or plasma are the most commonly used measurements of human selenium exposure. Blood is relatively easy to sample and contamination can be controlled. Studies on animals have shown that blood-selenium values are a good indication of selenium deficiency and excess. However, variations in selenium deficiency signs have been observed in animals with similar whole blood-selenium levels. This may be due to variations in the vitamin E content of the diet or exposure to other dietary or environmental variables. Several factors besides selenium intake have been identified that may affect blood-selenium content. Exposure to inorganic mercury or cadmium can lead to deposition of selenium in the blood attached to protein in combination with the metal. Human beings exposed to selenium in the form of selenomethionine or selenium-rich wheat or yeast, for 10 - 11 weeks, had higher blood-selenium levels than human beings exposed to the same amount of selenium in the form of selenite or selenate.

Blood can be fractionated and plasma-selenium and red blood cell-selenium can be measured separately. A recent study on the effects of a low-selenium diet on human beings indicated that a decrease in plasma-selenium content might occur before a decrease in red blood cell-selenium can be detected. These data suggest that it may be possible to use plasma-selenium levels to assess short-term selenium exposure but that red blood cell- and whole blood-selenium reflect long-term exposure.

9.3.4   Hair-selenium

Measurement of the selenium content in hair has been used in animals for the assessment of selenium status, with regard to both deficiency and excess. It has not been generally adopted for use in human beings, but, in special circumstances where external contamination can be excluded, hair-selenium content is useful for assessing selenium status.

9.3.5   Urine-selenium

The results of animal studies and a few human studies indicate that the urinary excretion of selenium can be useful for assessing very recent selenium exposure (i.e., within the past 24 h). However, determination of selenium in incomplete urine collections or expression of urinary-selenium per unit volume of urine cannot provide valid information about selenium exposure in the general population. An understanding of the reservations associated with this technique is necessary in its application and then it may only be useful in conjunction with other measurements discussed in this section.

9.3.6   Blood-glutathione peroxidase

There is no suitable, simple field method for assessing the selenium exposure of the general population. The Task Group concluded that approaches based on the glutathione peroxidase activity of blood components might provide the basis for a suitable screening method to detect low human exposure to selenium.

Blood-glutathione peroxidase activity is useful for detecting selenium deficiency, because it represents a functional form of selenium and its assay is more rapid than the measurement of selenium in blood.

However, the Task Group recognized that there are several difficulties and limitations associated with the determination of glutathione peroxidase activity. For example, the enzyme is not stable and hence its activity cannot be determined in samples stored for periods of time, whereas measurements of selenium can be made on stored samples. Moreover, large differences in blood-glutathione peroxidase activity have been reported from different laboratories using similar methods. The discrepancies can be so great that interlaboratory comparisons are impossible without suitable controls. Furthermore, results from New Zealand suggest that human blood-glutathione peroxidase activity may be useful in assessing selenium intake at low, but not necessarily at intermediate or high levels, because the activity of the enzyme is correlated with whole blood-selenium levels, only when the latter are less than 0.10 mg/litre. It is not known whether excessive selenium exposure can increase glutathione peroxidase activity in human blood, above normal values. Animal studies indicate that iron deficiency decreases blood glutathione peroxidase activity, thus possibly further confounding selenium status assessment by this method.

Human plasma contains glutathione peroxidase but its activity is very low and difficult to detect and haemolysis during sample collection should be excluded. Nevertheless, if these reservations are borne in mind, the technique is promising. Measurement of platelet-glutathione peroxidase is also a promising technique for measuring short-term changes in selenium status because of the short half-life of this blood component.

9.4 Levels of Dietary Selenium Exposure in the General Population

The selenium content of food is highly variable in relation to several factors. Levels of selenium in soil available for uptake by plants vary markedly in different locations and this is reflected in differences in the selenium contents of feeds and foodstuffs. Countries in which foodstuffs are shipped between regions tend to avoid extremes in dietary-selenium intake by averaging foodstuffs with high- and low-selenium contents but some differences in regional intakes are still observed.

Animal tissues usually do not have as high a selenium content as the plants in high-selenium areas or as low a selenium content as plants in low-selenium areas. This is probably because of the ability of animals to conserve selenium when it is in short supply, and to excrete it when an excess is present. This may protect consumers of animal products from extremes of selenium intake.

Available analytical data show that the levels of selenium typically found in foods, are in the range of 0.4 - 1.5 mg/kg in liver, kidney, and seafood; 0.1 - 0.4 mg/kg in muscle meats; from less than 0.1 to over 0.8 mg/kg in cereals or cereal products; less than 0.1 - 0.3 mg/kg in dairy products; and less than 0.1 mg/kg in most fruits and vegetables. The very large variation in the selenium contents of foodstuffs of the same type, depending on their origin, makes a food table approach to estimating dietary-selenium intake potentially misleading as discussed in section 9.3.2. More accurate intake estimates could be derived from assay data on samples of the food items actually being consumed (food-on-the-plate analysis).

The practice of supplementing the diets of livestock and poultry in low-selenium areas with selenite or selenate causes little increase in muscle-selenium content over levels encountered in animals raised in areas of adequate, but not excessive, selenium supply. With some exceptions, food processing and preparation generally do not cause major losses of selenium.

There are few data that indicate the chemical form that selenium takes in normal human foods. One available study of seleniferous wheat has shown that a significant fraction of the selenium is in the form of selenomethionine.

Various forms of selenium differ in their nutritional availability. The absorption of selenium from foods appears quite efficient (about 80%).

Daily selenium intake can be markedly influenced by food consumption patterns. For example, in many countries, consumption of large amounts of fish, kidney, or liver could raise an individual's selenium intake substantially above the highest daily intake shown in Table 8.

The greatest extremes in dietary-selenium intake have been reported in areas in which the diet was monotonous and consisted largely of locally-produced staple food.

The wide range of geographically-related selenium intake due to variations in the selenium contents of the diet is reflected in the wide range of selenium levels observed in human whole blood.

The extreme mean blood-selenium values reported for groups dependent on locally-grown foods of different selenium content ranged from 0.02 - 3.2 mg/litre (Table 11). Limited studies from New Zealand have shown that children and older individuals had lower whole blood-selenium levels, but it is not known whether these lower levels are due to low dietary intakes or to changes connected with growth and/or aging.

9.5   Evaluation of Health Risks - General Population

9.5.1   Predictive value of animal studies

The Task Group concluded that, in view of the still fragmentary data concerning the possible health effects of either deficient or excess selenium exposure on human beings, any evaluation of the human health effects of selenium exposure must take into account results arising from animal studies. For this purpose, these can be summarized as follows:

(a)   Selenium deficiency combined with concurrent low vitamin E status has resulted in deleterious effects in all animal species tested so far (mice, rats, chicks, ducks, swine, sheep, cattle, and monkeys). In rats, specific signs of selenium deficiency have been produced in animals fed diets adequate in vitamin E.

(b)   Acute and chronic selenium toxicity have been demonstrated in a wide variety of species, under a wide variety of conditions (section 7.1.2). In evaluating the toxic effects of different selenium compounds, the Task Group felt that it was useful to distinguish between effects that are strictly dependent on the selenium in the molecule and cannot be duplicated by, e.g., homologous compounds of sulfur, and compounds where the toxicity is, in principle, similar for homologues containing either selenium or sulfur. In addition, the Task Group recognized that the similarity of some of the effects of different selenium compounds could be related to the formation of certain common intermediates.

(c)   Dose-response relationships have been demonstrated in acute and chronic toxicity, as well as in selenium deficiency. Comparative studies have shown that acute toxicity is similar in parenteral and oral exposure (section 7.1.2.1), which is in good agreement with the recognized high absorbability of selenium compounds from the gastrointestinal tract (section 6.1.1).

(d)   As indicated in section 7.1.2.2, the borderline level of dietary selenium needed to cause growth depression, due to overt chronic selenium toxicity in rats, is in the range of 4 - 5 mg selenium/kg diet. However, the dietary level of selenium needed to cause chronic toxicity can be influenced by several environmental factors such as previous selenium intake (section 7.1.6).

(e)   The effects of various environmental factors on selenium dose-response relationships is even more dramatically illustrated in situations of low-selenium intake leading to deficiency. One of the primary factors influencing the nutritional requirements of animals for selenium is the vitamin E status. The minimum dietary level of selenium needed to prevent deficiency diseases in various animal species is in the range of 0.02 - 0.05 mg/kg diet.

(f)   Animal diseases associated with both high- and low-selenium intake have been reported in certain areas of the world in which the animals were consuming feeds primarily of local origin.

The possibility of a human disease due to selenium deficiency or excess should be looked for under similar conditions of restricted dietary consumption (monotonous diets based on locally grown foods). In addition, the Task Group concluded that, in view of the dependence of selenium deficiency or toxicity effects on various environmental factors, the possible involvement of selenium in human diseases of multifactorial etiopathogenesis deserves special attention.

9.5.2 <u>Studies on high-exposure effects in the general population</u>

Studies on this type of exposure are few and, as they are confined to subpopulations in high-selenium areas, some do not include comparison groups. Symptoms and signs of illness elicited are frequently mild and not clearly related to selenium. In practically all studies (section 8.1.1.1), nail pathology was reported and, in several studies, hair loss and increased dental decay. Some of the studies included reports of gastrointestinal disturbances and icteroid skin. Other possible causes of illness in these studies cannot be excluded and the dependence of severity of effects on gradation of selenium exposure cannot be evaluated. Examination of exposed individuals showed increased levels of selenium in the blood and urine.

Three studies on populations living in seleniferous areas and dependent on locally-produced food deserve particular attention. Examination of individuals exposed in these areas revealed highly increased levels of selenium in the blood and/or urine. A study from South Dakota reported various signs and symptoms in people with long-term overexposure to selenium, as revealed by elevated urinary-selenium excretion. In these areas, farm animals were affected by chronic selenium poisoning.

In a seleniferous zone of Venezuela (Villa Bruzual) 111 children (average blood-selenium level = 0.813 mg/litre) were studied and compared with 50 children (blood-selenium level = 0.355 mg/litre) not over-exposed to selenium (Caracas). The children from the seleniferous area had some loss of hair and some abnormalities of skin and nails; the authors noted some differences in socio-economic factors between the 2 groups.

In a recent report from China, past incidents of intoxication were described that were thought to be due to chronic selenium poisoning, with hair loss and nail pathology as the most common signs. No quantitative data are available from the period 1961-64, which was the time of peak prevalence of the intoxication. Recent diet samples collected from high-selenium areas with, and without, a history of this intoxication showed mean daily selenium intakes of about 5 and 0.75 mg, respectively (Table 7). Blood samples taken from the same areas had mean selenium levels of 3.2 and 0.44 mg/litre, respectively (Table 11).

The above studies of populations ingesting high selenium levels in Venezuela and China have not revealed abnormal serum transaminase activity in the subjects concerned. In indivi-

duals exposed to high levels of selenium in the recent
outbreak in the USA; there was no evidence of impaired liver
function, but loss of hair and pathological changes in the
nails were observed, as well as some other signs and symptoms
described in more detail in section 8.1.1.1. However, the
Task Group recognized that this is insufficient evidence to
exclude the occurrence of liver damage and recognized the need
for more thorough evaluation of the liver in persons with high
selenium exposure.

The Task group noted that the signs and symptoms of
selenium overexposure in human beings were not well defined.
However the Task Group was aware of 3 situations concerning
elevated selenium exposure, which could form the basis for
estimating a dose-response to high levels of selenium. In one
area of Enshi county in China, blood-selenium levels in adults
of 0.44 mg/litre were associated with no reported effects. In
Venezuelan children with blood-selenium levels of 0.813
mg/litre, hair and nail changes of an unspecified nature and
incidence were reported. In another area of Enshi county in
China, definite hair and nail changes and symptoms and signs
consistent with neuropathy were reported in adults with blood
levels of 3.2 mg/litre. The Task Group, however, could not be
certain of how signs and symptoms were searched for, how their
absence was ascertained, and how the control populations were
selected in these studies.

9.5.3   Studies on low-exposure effects in the general
        population

Very low blood-selenium levels observed in human subjects
(section 8.2.4), approach those observed in animals with
selenium responsive diseases. One woman from a general
population known to have a low exposure to selenium and
maintained on total parenteral nutrition, which provided less
than 1 µg selenium/day, had muscular pain and dysfunction
that responded to selenium supplementation. However, similar
symptoms have not been observed in other patients with similar
low blood-selenium levels. Also, in another study, no
muscular symptomatology was reported in children fed
exclusively special medical diets containing very low levels
of selenium, for long periods of time. On the other hand, the
Task Group assigned a high significance to recent reports
describing an association between Keshan Disease (section
9.5.4.1) and poor selenium status as indicated by low selenium
content of grain, low blood and hair levels of selenium, low
blood-glutathione peroxidase activity, as well as a positive
response to an intervention study with sodium selenite. As
discussed in section 9.5.1, the results of several
experimental animal studies indicate that many different

factors may contribute to the development of selenium deficiency diseases. These studies point to the conclusion that a lack of selenium may be only one of several causative factors responsible for the occurrence of certain diseases of complex multiple etiopathogenesis. Such a conclusion was also fully recognized in the reports from China on the Keshan Disease.

9.5.4 Evaluation of the involvement of selenium in human diseases of multiple etiopathogenesis

9.5.4.1 Keshan disease

A suitable animal model of the Keshan disease is not available, even though heart damage is a feature of combined vitamin E and selenium deficiency in several species of animals.

The ecological evidence strongly favours a relationship between low selenium status and the incidence of Keshan disease. Such evidence includes low blood-, hair-, and urine-selenium levels in the affected areas as well as a low selenium content in the staple foods raised and consumed in the affected areas. A randomized intervention trial carried out in China showed that children who received sodium selenite had a lower incidence of Keshan disease than those who received a placebo. However, in this randomized trial there may have been an underlying trend in the placebo group towards a decline in the incidence of the disease. Also, information on the effectiveness of the randomization was not available and the incidence rates were not adjusted for age. The results of the much larger but non-randomized intervention trial involving at least a quarter of a million children on a yearly basis indicated a clear-cut beneficial effect of sodium selenite in the prevention of the Keshan disease. Thus, the Task Group concluded that Keshan disease is a condition in human beings that is related to low selenium status but that additional research is needed to clarify the role of other factors in the etiopathogenesis of the disease.

The Task Group also recognized that in section 8.2.3.1 this disease was used as the basis for estimating a minimal human nutritional selenium requirement, suggested by the authors to be 19 and 14 µg/day for men and women, respectively.

9.5.4.2 Kashin-Beck disease

An additional disease found in certain low-selenium areas of China is an endemic osteoarthropathy known as Kashin-Beck disease that occurs mainly in children. There is some

evidence linking low selenium status with the incidence of
Kashin-Beck disease, but the Task Group concluded that
additional research, particularly in relation to intervention
studies, is required before a definite statement can be made
concerning the role of selenium in the etiology of the disease.

9.5.4.3 Ischaemic heart disease

The Task Group was aware of studies that examined the
possibility of an association between low selenium status and
myocardial ischaemia and atherosclerosis. However, because of
the limited data available, the Task Group was not prepared to
come to any conclusion regarding the role of selenium in
ischaemic heart disease.

9.5.4.4 Studies on the involvement of selenium in cancer

(a) Carcinogenicity studies

The Task Group concluded that the studies described in
section 7.7 suggesting a carcinogenic effect of selenite or
selenate were invalid, because, in one study, the results were
statistically insignificant and, in the other studies, there
were no controls or systematic pathological examinations. In
another systematic study, no differences were observed in the
incidence of tumours in rats surviving 2 years or longer and
exposed to 0.5 - 2.0 mg selenium/kg diet.
The Task Group was aware of one study that demonstrated
the carcinogenicity of ethyl selenac (selenium
diethyldithiocarbamate) and another that showed the
carcinogenicity of selenium sulfide, given by gavage. The
latter result is of possible interest both as regards human
dermal exposure to selenium sulfide and/or its possible
formation within the body, but the Task Group was not aware of
any studies demonstrating carcinogenic effects of other
selenium compounds.
The Task Group felt that the evaluation published by IARC
in 1975 was still valid in concluding that the available
animal data were insufficient to allow an evaluation of the
carcinogenicity of selenium compounds (IARC, 1975).

(b) Experimental evidence of anticarcinogenic effects of selenium compounds

The results of studies on laboratory animals provide
evidence of a preventive effect of selenium dioxide, sodium
selenite, or "selenized" yeast, given in the food or drinking-
water, against chemically-induced cancers or certain
spontaneous, presumably viral-induced cancers (section 7.7).

Generally, the level required to demonstrate such effects ranges from 1 - 6 mg/kg food or per litre drinking-water and thus is considerably in excess of the animal's nutritional needs. In one report, selenium dioxide at 0.1 mg selenium/litre water had some beneficial effect against spontaneous mammary tumours in mice, but, in this case, the diet itself contained 0.45 mg selenium/kg. Other studies have shown the beneficial effects of sodium selenide when applied concomitantly to the skin with certain carcinogens. In rats fed diets high in polyunsaturated fats, selenium deficiency increased the incidence of mammary tumours, after treatment with dimethylbenz[α]anthracene. Therefore, the Task Group concluded that pharmacological levels of selenium compounds exhibit in many cases a favourable influence against the development of cancer in various animal model systems. However, the Task Group was aware of one study in which administration of high levels of selenium was associated with an increased incidence of chemically-induced cancer.

(c) <u>Epidemiological evidence regarding the possible anti-carcinogenic effects of selenium in human beings</u>

The Task Group acknowledged that apparent negative correlations have been drawn between cancer death rates and certain general population characteristics, such as blood-selenium levels or the average level of selenium in diets in specific geographical zones (section 8.2.6.1). However, these apparent correlations have not been consistent with specific cancer sites and are subject to ecological fallacy. Moreover, such epidemiological data are subject to question regarding the adequacy of sampling, the interpretation of blood-selenium data, and the precision of the estimates of dietary-selenium intake in various countries and different geographical areas within the USA.

Of the 7 case-control studies examined by the Task Group, no consistent association was apparent between low selenium status and risk of cancer. For a given cancer site, there were few replicate studies. On the majority of occasions, no association was found between low selenium status and cancer risk. In fact, in at least one situation a positive correlation was found. In several countries, low blood- or plasma-selenium levels were found in patients with several different types of cancer. However, the effect of the malignant disease itself in lowering blood-selenium levels (e.g., by decreasing absorption and/or worsening the generally debilitated nutritional state of the patient) cannot be excluded as a partial or total explanation for these results.

The Task Group was aware of 3 studies in which blood samples had been taken for other purposes prior to the diagnosis of any cancer. Some time later, the samples were retrieved from storage banks and analysed for selenium. All 3 studies showed a consistent inverse association between the prediagnostic serum-selenium level and the risk of cancer. This association between low serum-selenium level and cancer risk was observed only in smokers, only in males, and in one study only in blacks but not in whites. In one study, the association between low selenium level and cancer risk showed some site specificity, but the second and third studies were unable to confirm or rebut this because of inadequate sample size.

The Task Group was aware that in many studies showing a protective effect of selenium against cancer in animals pharmacological doses of selenium compounds were used. Therefore, there were difficulties in relating the animal and human studies, but the Task Group was aware of an intervention trial being carried out in China which it is hoped will enable a more precise evaluation of the association between selenium status and human cancer risk in the future.

Thus, the Task Group concluded that existing data are insufficient to determine whether the level of selenium intake is indeed correlated with the incidence of cancer in man.

9.5.4.5 Caries

Although an association between high selenium intake and dental caries has been reported in at least one animal study, there is no clear-cut evidence of such an association in man. The Task Group concluded that difficulties associated with the interpretation of urinary-selenium excretion, expressed as mass-concentration per volume and not expressed as total daily urinary excretion of selenium (section 9.3.5), as well as the inability to exclude interference by other environmental factors such as fluoride, precluded any significant conclusions. A better understanding of the impact of selenium on the incidence of caries will require a more comprehensive estimation of selenium exposure and of the other confounding factors in the respective subpopulations under study.

9.5.4.6 Health effects related to reproduction

The Task Group concluded that at present there is no evidence that selenium has significant effects on reproductive function in man.

9.6 Occupational Exposure

In contrast to general population exposure, occupational exposure usually occurs through direct contact and/or through inhalation, i.e., dermal and respiratory exposure predominate. Exposure to selenium and its compounds occurs in primary industries, i.e., those that extract, mine, treat, or process selenium-bearing minerals, e.g., copper, zinc, or lead ores and in secondary industries, i.e., those that use selenium in manufacturing processes. The physical and chemical form of selenium under these circumstances varies and is determined by the industrial processes. The Task Group recognized that industrial exposure has not been adequately studied with respect to levels of exposure and there is a need for such studies to be carried out.

For acute occupational selenium exposures, the effects vary according to to the chemical form of selenium. In contrast to elemental selenium, which does not appear to be toxic unless oxidized or reduced, hydrogen selenide and selenium dioxide are highly toxic, causing irritation of the respiratory tract, which may be followed by pulmonary oedema.

The Task Group recognized that evaluation of long-term occupational exposure to selenium must take into account other dietary and environmental substances because of their recognized or potential interactions with selenium. An additional factor that needs to be considered is the fact that preventive measures usually result in the removal of exposed workers with the appearance of the first sign of selenium over-exposure (garlic-like odour). Monitoring of selenium in the urine is also used to identify those who should be removed from further overexposure.

The Task Group concluded that no studies were available on the dose-response relationship with regard to occupational exposure to selenium.

REFERENCES

ABDULLA, M., KOLAR, K., & SVENSSON, S. (1979) Selenium. Scand. J. Gastroenterol., 14(suppl. 52): 181-184.

ABDULLAEV, G.M. (1976) [Selenium levels in healthy subjects and those with certain haematological diseases.] In: Selen v biologii, Vol. 1, pp. 136-139, Baku, Elm Publishing House (in Russian).

ABDULLAEV, G.B., GADZHIEVA, N.A., GASANOV, G.G., DZHAFAROV, A.I., & PERELYGIN, V.V. (1972) [Selenium and vision,] Baku, Elm Publishing House, 50 pp (in Russian).

ABDULLAEV, G.B., GADZHIEVA, N.A., GASANOV, G.G., MAMEDOVA, S.A., DMITRENKO, A.I., GULIEVA, L.I., & RODIONOV, V.P. (1974a) [The effect of selenium compounds on electroretinogram under different experimental conditions.] In: [Selen v biologii,] Baku, Elm Publishing House, pp. 11-22 (in Russian).

ABDULLAEV, G.B., MAMEDOV, SH.V., DZHAFAROV, A.I., PERELYGIN, V.V., & MAGOMEDOV, N.M. (1974b) [Possible regulation of free radicals in the retina by selenium compounds.] In: [Selen v biologii,] Baku, Elm Publishing House, pp. 54-57 (in Russian).

ABU-ERREISH, G.M. (1967) On the nature of some selenium losses from soils and waters, Brookings, South Dakota, South Dakota State University (MS Thesis).

ABU-ERREISH, G.M., WHITEHEAD, E.I., & OLSON, O.E. (1968) Evolution of volatile selenium from soils. Soil Sci., 106: 415-420.

ABUTALYBOV, M.G., VEZIROVA, N.B., FATALIEVA, C.M., KHALILOV, E.KH., & MUSAEV, S.G. (1976) [Selenium content in some bean plants of Azerbajdzhan.] In: [Selen v biologii,] Vol. 2, Baku, Elm Publishing House, pp. 140-142 (in Russian).

ACGIH (1971) Documentation of the threshold limit values for substances in workroom air, 3rd ed., Cincinnati, Ohio, American Conference of Governmental Industrial Hygienists, pp. 224-225.

AKESSON, B. (1985) Plasma selenium in patients with abnormal plasma protein patterns. In: Proceedings of the XIII International Congress of Nutrition, p. 152 (Abstract).

ALEXANDER, A.R., WHANGER, P.D., & MILLER, L.T. (1983) Bioavailability to rats of selenium in various tuna and wheat products. J. Nutr., 113(1): 196-204.

ALCINO, J.F. & KOWALD, J.A. (1973) Analytical methods. In: Klayman, D.L. & Gunther, W.H.H., ed. Organic selenium compounds, their chemistry and biology, New York, John Wiley and Sons, pp. 1049-1081.

ALLAWAY, W.H. (1972) An overview of distribution patterns of trace elements in soils and plants. Ann. NY Acad. Sci., 199: 17-25.

ALLAWAY, W.H. (1973) Selenium in the food chain. The Cornell Veterinarian, 63: 151-170.

ALLAWAY, W.H. (1978) Perspectives on trace elements in soil and human health. In: Hemphill, D.D., ed. Trace substances in environmental health. XII, Columbia, Missouri, University of Missouri Press, pp. 3-10.

ALLAWAY, W.H. & CARY, E.E. (1964) Determination of submicrogram amounts of selenium in biological materials. Anal. Chem., 36: 1359-1362.

ALLAWAY, W.H., MOORE, D.P., OLDFIELD, J.E., & MUTH, O.H. (1966) Movement of physiological levels of selenium from soils through plants to animals. J. Nutr., 88: 411-418.

ALLAWAY, W.H., CARY, E.E., & EHLIG, C.F. (1967) The cycling of low levels of selenium in soils, plants, and animals. In: Muth, O.H., Oldfield, J.E., & Weswig, P.H., ed. Selenium in biomedicine, Westport, Connecticut, The AVI Publishing Co., Inc, pp. 273-296.

ALLAWAY, W.H., KUBOTA, J., LOSEE, F., & ROTH, M. (1968) Selenium, molybdenum, and vanadium in human blood. Arch. environ. Health, 16: 342-348.

AMOR, A.J. & PRINGLE, P. (1945) A review of selenium as an industrial hazard. Bull. Hyg., 20: 239-241.

ANCIZAR-SORDO, J. (1947) Occurrence of selenium in soils and plants of Columbia, South America. Soil Sci., 63: 437.

ANDERSON, R.A. & POLANSKY, M. (1981) Dietary chromium deficiency. Effect on sperm count and fertility in rats. Biol. Trace Elem. Res., 3: 1-5.

ANDREWS, R.W. & JOHNSON, D.C. (1976) Determination of selenium (IV) by anodic stripping voltammetry in flow system with ion exchange separation. Anal. Chem., 48: 1056-1060.

ANONYMOUS (1962) Selenium poisons Indians. Sci. News Lett., 81: 254.

ANONYMOUS (1975) New links in the selenium cycle. Agric. Res., 24: 6-7.

ANONYMOUS (1984) Washington DC, US Food and Drug Administration, pp. 19 (FDA Bulletin No. 14).

ANUNDI, I., STAHL, A., & HOGBERG, J. (1984) Chem.-biol. Interact., 50: 277.

AOAC (1975) Official methods of analysis, 12th ed., Washington DC, Association of Official Analytical Chemists.

ARTHUR, D. (1972) Selenium content of Canadian foods. Can. Inst. Food Sci. Technol. J., 5: 165-169.

ASHER, C.J., BUTLER, G.W., & PETERSON, P.J. (1977) Selenium transport in root systems of tomato. J. exp. Bot., 28: 279-291.

AWASTHI, Y.C., BEUTLER, E., & SRIVASTAVA, S.K. (1975) Purification and properties of human erythrocyte glutathione peroxidase. J. biol. Chem., 250: 5144-5149.

BACHAREV, V.D., BOCHAROVA, M.A., & SHOSTAK, V.I. (1975) [On the effect of selenium on the light perception of the eye.] Fiziol. Zhurn., 61(1): 150-153 (in Russian).

BAI, J., WU, S.Q., GE, K.Y., DENG, X.J., & SU, C.Q. (1980) The combined effect of selenium deficiency and viral infection on the myocardium of mice (preliminary study). Acta Acad. Med. Sinicae, 2: 29-31.

BAIRD, R.B., POURIAN, S., & GABRIELIAN, S.M. (1972) Determination of trace amounts of selenium in waste waters by carbon rod atomization. Anal. Chem., 44: 1887-1889.

BARBEZAT, G.O., CASEY, C.E., REASBECK, P.G., ROBINSON, M.F., & THOMSON, C.D. (1984) Selenium. In: Rosenberg, I. & Solomons, N.W., ed. Absorption and malabsorption of mineral nutrients, New York, Alan R. Liss., pp. 231-258.

BARRETTE, M., LAMOUREUX, G., LEBEL, E., LECOMTE, R., PARADIS, P., & MONARO, S. (1976) Trace element analysis of freeze-dried blood serum by proton and α-induced X-rays. Nucl. Instrum. Methods, 134: 189-196.

BEARSE, R.C., CLOSE, D.A., MALANIFY, J.J., & UMBARGER, C.J. (1974) Elemental analysis of whole blood using proton-induced X-ray emission. Anal. Chem., 46: 499-503.

BEATH, O.A., EPPSON, H.F., & GILBERT, C.S. (1935). Selenium and other toxic minerals in soils and vegetation., Wyoming Experimental Station, 55 pp (Bulletin No. 206).

BEHNE, D. & HOFER-BOSSE, T. (1984) Effects of a low selenium status on the distribution and retention of selenium in the rat. J. Nutr., 114: 1289-1296.

BEHNE, D. & WOLTERS, W. (1979) Selenium content and gluta-thione peroxidase activity in the plasma and erythrocytes of non-pregnant and pregnant women. J. clin. Chem. clin. Biochem., 17: 133-135.

BEIJE, B., ONFELT, A., & OLSSON, U. (1984) Influence of dietary selenium on the mutagenic activity of perfusate and bile from rat liver, perfused with 1,1-dimethylhydrazine. Mutat. Res., 130: 121-126.

BERTINE, K.K. & GOLDBERG, E.D. (1971) Fossil fuel combustion and the major sedimentary cycle. Science, 173: 233-235.

BESBRIS, H.J., WORTZMAN, M.S., & COHEN, A.M. (1982) Effect of dietary selenium on the metabolism and excretion of 2-acetylaminofluorene in the rat. J. Toxicol. environ. Health, 9: 63-76.

BHUYAN, K.C., BHUYAN, D.K., & PODOS, S.M. (1981) Selenium-induced cataract: biochemical mechanism. In: Spallholz, J.E., Martin, J.L., & Ganther, H.E., ed. Selenium in biology and medicine, Westport, Connecticut, Avi Publishing Company, pp. 403-412.

BIERI, J.G. & AHMAD, K. (1976) Selenium content of Bangladeshi rice by chemical and biological assay. J. agric. food Chem., 24: 1073-1074.

BIRT, D.F., JULINS, A.D., & POUR, P.M. (1984) Increased pancreatic carcinogenesis in Syrian hamsters fed high selenium diets. Proc. Am. Assoc. Cancer Res., 25: 133.

BLADES, M.W., DALZIEL, J.A., & ELSON, C.M. (1976) Cathodic stripping voltammetry of nanogram amounts of selenium in biological material. J. Assoc. Off. Anal. Chem., 59: 1234-1239.

BLOTCKY, A.J., SULLIVAN, J.F., SHUMAN, M.S., WOODWARD, G.P., VOORS, A.W., & JOHNSON, W.D. (1976) Selenium levels in liver and kidney. In: Hemphill, D.D., ed. Trace substances in environmental health. X, Columbia, Missouri, University of Missouri Press, pp. 97-103.

BONARD, E.C. & KORALINK, K.D. (1958) Intoxication aiguë par vapeurs d'hydrogene selenide. Praxis, 47(i): 533-534.

BOUND, G.P. & FORBES, S. (1978) Differential-pulse polarography of selenium (IV) in the presence of metal ions. Analyst, 103: 176-179.

BOWEN, H.J.M. & CAWSE, P.A. (1963) The determination of selenium in biological material by radioactivation. Analyst, 88: 721-726.

BOWEN, W.H. (1972) The effects of selenium and vanadium on caries activity in monkeys (M. irus). J. Irish Dent. Assoc., 18: 83-89.

BRADY, P.S., KU, P.K., & ULLREY, D.E. (1978) Lack of effect of selenium supplementation on the response of the equine erythrocyte glutathione system and plasma enzymes to exercise. J. anal. Sci., 47: 492-496.

BRADY, P.S., BRADY, L.J., & ULLREY, D.E. (1979) Selenium, vitamin E, and the response to swimming stress in the rat. J. Nutr., 109: 1103-1109.

BRITTON, J.L., SHEARER, T.R., & DE SART, D.J. (1980) Cariostasis by moderate doses of selenium in the rat model. Arch. environ. Health, 35: 74-76.

BRODIE, K.G. (1979) Analysis of arsenic and other trace elements by vapor generation. Am. Lab. (June issue): 58-66.

BROGHAMER, W.L., MCCONNELL, K.P., & BLOTCKY, A.L. (1976) Relationship between serum selenium levels and patients with carcinoma. Cancer, 37: 1384-1388.

BROGHAMER, W.L., MCCONNELL, K.P., GRIMALDI, M., & BLOTCKY, A.J. (1978) Serum selenium and reticuloendothelial tumours. Cancer, 41: 1462-1466.

BROWN, D.G. & BURK, R.F. (1973) Selenium retention in tissues and sperm of rats fed a torula yeast diet. J. Nutr., 103: 102-108.

BROWN, D.G., BURK, R.F., SEELY, R.J., & KIKER, K.W. (1972) Effect of dietary selenium on the gastrointestinal absorption of $^{75}SeO_3$ in the rat. Int. J. Vit. nutr. Res., 42: 588-591.

BROWN, M.W. & WATKINSON, J.H. (1977) An automated fluorimetric method for the determination of nanogram quantities of selenium. Anal. Chim. Acta, 89: 29-35.

BRUNE, D., SAMSAHL, K., & WESTER, P.O. (1966) A comparison between the amounts of As, Au, Br, Cu, Fe, Mo, Se, and Zn in normal and uraemic human whole blood by means of neutron activation analysis. Clin. Chim. Acta, 13: 285-291.

BUCHAN, R.F. (1947) Industrial selenosis. Occup. Med., 3: 439-456.

BUNCE, G.E., HESS, J.L., GURLEY, R., BATRA, R., & TARNAWSKA, E. (1985) Lens energy metabolism and Ca content in selenite induced cataract. In: Mills, C.F., Bremner, I., & Chesters, J.K., ed. Trace elements in man and animals - TEMA 5, Slough, Commonwealth Agricultural Bureaux, pp. 250-254.

BURCHANOV, A.I. (1972) [On regulation of complex chemical dusts in rare metals production.] Vopr. Gig. Trud. Profzabol. Kazajenade, Kazah NII Gig. Trud. Profazabol., pp. 72-74 (in Russian).

BURCHANOV, A.I. & ZHAKENOVA, R.K. (1973) [Pulmonary changes following intratracheal administration of elementary selenium to white rats.] Zavavoohr. Kazah., 3: 52-53 (in Russian).

BURCHANOV, A.I., SALEHOV, M.I., DOROFEEVA, O.N., & ZHAKENOVA, R.K. (1969) [On the effects of metal selenium dust on the organism.] In: [Proceedings of a Concluding Scientific Conference on Labour Hygiene and Occupational Disease, 21-22 October, 1969,] Karaganda, Kazah., NII Gig. Trud. Profzabol, pp. 77-80 (in Russian).

BURK, R.F. (1976) Selenium in man. In: Prasad, A.S., ed. Trace elements in human health and disease. II. Essential and toxic elements, New York, Academic Press, pp. 105-133.

BURK, R.F. & CORREIA, M.A. (1977) Accelerated hepatic haem catabolism in the selenium-deficient rat. Biochem. J., 168: 105-111.

BURK, R.F. & LANE, J.M. (1979) Ethane production and liver necrosis in rats after administration of drugs and other chemicals. Toxicol. appl. Pharmacol., 50: 467-478.

BURK, R.F & MASTERS, B.S.S. (1975) Some effects of selenium deficiency on the hepatic microsomal cytochrome P-450 system in the rat. Arch. Biochem. Biophys., 170: 124-131.

BURK, R.F., Jr, PEARSON, W.N., WOOD, R.P., II, & VITERI, F. (1967) Blood selenium levels and in vitro red blood cell uptake of ^{75}Se in kwashiorkor. Am. J. clin. Nutr., 20: 723-733.

BURK, R.F., BROWN, D.G., SEELY, R.J., & SCAIEF, C.C., III (1972) Influence of dietary and injected selenium on whole-body retention, route of excretion, and tissue retention of ^{75}SeO$_3$ $^{2-}$ in the rat. J. Nutr., 102: 1049-1055.

BURK, R.F., SEELEY, R.J., & KIKER, K.W. (1973) Selenium: dietary threshold for urinary excretion in the rat. In: Proceedings of the Society of Experimental and Biological Medicine, Vol. 142, pp. 214-216.

BURK, R.F., MACKINNON, A.M., & SIMON, F.R. (1974) Selenium and hepatic microsomal hemoproteins. Biochem. Biophys. Res. Commun., 56: 431-436.

BURK, R.F., NISHIKI, K., LAWRENCE, R.A., & CHANCE, B. (1978) Peroxide removal by selenium-dependent and selenium independent glutathione peroxidases in hemoglobin-free perfused rat liver. J. biol. Chem., 253: 43-46.

BURK, R.F., LAWRENCE, R.A., & CORREIA, M.A. (1980a) Sex differences in biochemical manifestations of selenium deficiency in rat liver with special reference to heme metabolism. Biochem. Pharmacol., 29: 39-42.

BURK, R.F., LAWRENCE, R.A., & LANE, J.M. (1980b) Liver necrosis and lipid peroxidation in the rat as the result of paraquat and diquat administration. J. clin. Invest., 65: 1024-1031.

BUS, J.S., AUST, S.D., & GIBSON, J.E. (1974) Superoxide- and singlet oxygen-catalyzed lipid peroxidation as a possible mechanism for paraquat (methyl viologen) toxicity. Biochem. Biophys. Res. Commun., 58: 749-755.

BUTLER, G.W. & PETERSON, P.J. (1961) Aspects of the faecal excretion of selenium by sheep. N.Z. J. agric. Res., 4: 484-491.

BUTTNER, W. (1963) Action of trace elements on the metabolism of fluoride. J. dent. Res., 42: 453-460.

BYARD, J.L. (1969) Trimethyl selenide: a urinary metabolite of selenite. Arch. Biochem. Biophys., 130: 556-560.

CADELL, P.B. & COUSINS, F.B. (1960) Urinary selenium and dental caries. Nature (Lond.), 185: 863-864.

CAGEN, S.Z. & GIBSON, J.E. (1977) Liver damage following paraquat in selenium-deficient and diethyl maleate-pretreated mice. Toxicol. appl. Pharmacol., 40: 193-200.

CALVIN, H.I. (1978) Selective incorporation of selenium-75 into a polypeptide of the rat sperm tail. J. exp. Zool., 204: 445-452.

CANNELLA, J.M. (1976) Surveillance of employees in selenium alloying operations. In: Proceedings of the Sympoisium on Selenium-Tellurium in the Environment, Pittsburgh, Pennsylvania, Industrial Health Foundation, pp. 343-348.

CANTOR, A.H. & SCOTT, M.L. (1974) The effect of selenium in the hen's diet on egg production, hatchability, performance of progeny, and selenium concentration in eggs. Poultry Sci., 53: 1870-1880.

CANTOR, A.H. & SCOTT, M.L. (1975) Influence of dietary selenium on tissue selenium levels in turkeys. Poultry Sci., 54: 262-265.

CANTOR, A.H., SCOTT, M.L., & NOGUCHI, T. (1975a) Biological availability of selenium in feedstuffs and selenium compounds for prevention of exudative diathesis in chicks. J. Nutr., 105: 96-105.

CANTOR, A.H., LANGEVIN, M.L., NOGUCHI, T., & SCOTT, M.L. (1975b) Efficacy of selenium in selenium compounds and feedstuffs for prevention of pancreatic fibrosis in chicks. J. Nutr., 105(1): 106-111.

CANTOR, A.H., MOORHEAD, P.D., & BROWN, K.I. (1978) Influence of dietary selenium upon reproductive performance of male and female breeder turkeys. Poultry Sci., 57: 1337-1345.

CAPPON, C.J. & SMITH, J.C. (1982) Chemical form and distribution of mercury and selenium in canned tuna. J. appl. Toxicol., 2: 181-189.

CARNRICK, G.R., MANNING, D.C., & SLAVIN, W. (1983) Determination of selenium in biological materials with platform furnace atomic absorption spectroscopy and Zeeman background correction. Analyst, 108: 1297-1312.

CARTER, D.L., ROBBINS, C.W., & BROWN, M.J. (1972) Effect of phosphorus fertilization on the selenium concentration in alfalfa (Medicago sativa). Soil Sci. Soc. Am. Proc., 36: 624-628.

CARY, E.E., ALLAWAY, W.H., & MILLER, M. (1973) Utilization of different forms of dietary selenium. J. anal. Sci., 36: 285-292.

CASEY, C.E., GUTHINE, B.E., FREUD, G.M., & ROBINSON, M.F. (1982) Selenium in human tissues from New Zealand. Arch. environ. Health., 37: 133-135.

CAYGILL, C.P.J. & DIPLOCK, A.T. (1973) The dependence on dietary selenium and vitamin E of oxidant-labile liver microsomal non-haem iron. FEBS Lett., 33: 172-176.

CHAN, C.C.Y. (1976) Improvement in the fluorimetric determination of selenium in plant materials with 2,3-diaminonaphthalene. Anal. Chim. Acta, 82: 213-215.

CHANSLER, M.W., MORRIS, V.C., & LEVANDER, O.A. (1983) Bioavailability to rats of selenium in Brazil nuts and mushrooms. Fed. Proc., 42(4): 927.

CHATTERJEE, M. & BANERJEE, M.R. (1982) Selenium mediated dose-inhibition of 7-12-dimethylbenz(α)anthracene-induced transformation of mammary cells in organ culture. Cancer Lett., 17: 187-195.

CHAU, Y.K., WONG, P.T.S., SILVERBERG, B.A., LUXON, P.L., & BENGERT, G.A. (1976) Methylation of selenium in the aquatic environment. Science, 192: 1130-1131.

CHAVEZ, J.F. (1966) Studies on the toxicity of a sample of Brazil nuts with a high content of selenium. Bol. Soc. Quim. Peru, 32: 195-203.

CHEN, R.W., WHANGER, P.D., & WESWIG, P.H. (1975) Selenium-induced redistribution of cadmium binding to tissue proteins: a possible mechanism of protection against cadmium toxicity. Bioinorgan. Chem., 4: 125-133.

CHEN, X., YANG, G., CHEN, J., CHEN, X., WEN, Z., & GE, K. (1980) Studies on the relations of selenium and Keshan disease. Biol. Trace Elem. Res., 2 91-107.

CHERRY, D.S. & GUTHRIE, R.K. (1977) Toxic metals in surface waters from coal ash. Water Resour. Bull., 13: 1227-1236.

CHUNG, A. & MAINES, M.D. (1981) Effect of selenium on glutathione metabolism. Induction of -glutamylcysteine synthetase and glutathione reductase in the rat liver. Biochem. Pharmacol., 30(23): 3217-3223.

CLARK, L.C. (1985) The epidemiology of selenium and cancer. Fed. Proc., 44(9): 2584-2589.

CLAYCOMB, C.K., SUMMERS, G.W., & JUMP, E.B. (1965) Effect of dietary selenium on dental caries in Sprague Dawley rats. J. dent. Res., 44: 826.

CLAYTON, C.C. & BAUMANN, C.A. (1949) Diet and azo dye tumors: effect of diet during a period when the dye is not fed. Cancer Res., 9: 575-582.

CLINTON, M., Jr (1947) Selenium fume exposure. J. ind. Hyg. Toxicol., 29: 225-226.

COMBS, G.F., Jr & SCOTT, M.L. (1975) Polychlorinated biphenyl-stimulated selenium deficiency in the chick. Poultry. Sci., 54: 1152-1158.

COMBS, G.F., Jr, CANTOR, A.H., & SCOTT, M.L. (1975) Effects of dietary polychlorinated biphenyls on vitamin E and selenium nutrition in the chick. Poultry Sci., 54: 1143-1152.

CONE, J.E., MARTIN DEL RIO, R., DAVIS, J.N., & STADTMAN, T.C. (1976) Chemical characterization of the selenoprotein component of clostridial glycine reductase: identification of selenocysteine as the organoselenium moiety. Proc. Natl Acad. Sci. (USA), 73: 2659-2663.

COOPER, W.C. (1967) Selenium toxicity in man. In: Muth, O.H., Oldfield, J.E., & Weswig, P.H., ed. Selenium in biomedicine, Westport, Connecticut, The AVI Publishing Co., Inc, pp. 185-199.

COOPER, W.C. (1974) Analytical chemistry of selenium. In: Zingaro, R.A. & Cooper, W.C., ed. Selenium, New York, Van Nostrand Reinhold Co., pp. 615-653.

COOPER, W.C. & GLOVER, J.R. (1974) The toxicology of selenium and its compounds. III. In: Zingaro, R.A. & Cooper, W.C., ed. Selenium, New York, Van Nostrand Reinhold Co., pp. 654-674.

COOPER, W.C., BENNETT, K.G., & CROXTON, F.C. (1974) The history, occurrence, and properties of selenium. In: Zingaro, R.A. & Cooper, W.C., ed. Selenium, New York, Van Nostrand Reinhold Co., pp. 1-31.

CORREIA, M.A. & BURK, R.F. (1976) Hepatic heme metabolism in selenium-deficient rats: effect of phenobarbital. Arch. Biochem. Biophys., 177: 642-644.

CORREIA, M.A. & BURK, R.F. (1978) Rapid stimulation of hepatic microsomal heme oxygenase in selenium-deficient rats: an effect of phenobarbital. J. biol. Chem., 253: 6203-6210.

COWGILL, U.M. (1974) Trace elements and the birth rate in the continental United States. In: Hemphill, D.D., ed. Trace substances in environmental health. VIII, Columbia, Missouri, University of Missouri Press, pp. 15-21.

COWGILL, U.M. (1976) Selenium and human fertility. In: Proceedings of the Symposium on Selenium and Tellurium in the Environment, Pittsburg, Pennsylvania, Industrial Health Foundation, pp. 300-315.

CREGER, C.R., MITCHELL, R.H., ATKINSON, R.L., FERGUSON, T.M., REID, B.L., & COUCH, J.R. (1960) Vitamin E activity of selenium in turkey hatchability. Poultry Sci., 39: 59-63.

CRYSTAL, R.G. (1973) Elemental selenium: structure and properties. In: Klayman, D.L. & Gunther, W.H.H., ed. Organic selenium compounds: their chemistry and biology, New York, John Wiley and Sons, Inc, pp. 13-27.

CSALLANY, A.S., SU, L.-C., & MENKEN, B.Z. (1984) Effect of selenite, vitamin E, and n,n'-diphenyl-p-phenylenediamine on liver organic solvent-soluble lipofuscin pigments in mice. J. Nutr., 114: 1582-1587.

CUKOR, P., WALZCYK, J., & LOTT, P.F. (1964) The application of isotopic dilution analysis to the fluorimetric determination of selenium in plant material. Anal. Chim. Acta, 30: 473-482.

CUMMINS, L.M. & KIMURA, E.T. (1971) Safety evaluation of selenium sulfide antidandruff shampoos. Toxicol. appl. Pharmacol., 20: 89-96.

CUMMINS, L.M. & MARTIN, J.L. (1967) Are selenocystine and selenomethionine synthesized in vivo from sodium selenite in mammals? Biochemistry, 6: 3162-3168.

CZERNIEJEWSKI, C.P., SHANK, C.W., BECHTEL, W.G., & BRADLEY, W.B. (1964) The minerals of wheat, flour, and bread. Cereal Chem., 41: 65-72.

D'HONDT, P., LIEVENS, P., VERSIECK, J., & HOSTE, J. (1977) Determination of trace elements in animal and human muscle by semi-automated radiochemical neutron activation analysis. Radiochem. Radioanal. Lett., 31: 231-240.

DAMS, R. & DE JONGE, J. (1976) Chemical composition of Swiss aerosols from the Jungfraujoch. Atmos. Environ., 10: 1079-1084.

DE JONG, D., MORSE, R.A., GUTENMANN, W.H., & LISK, D.J. (1977) Selenium in pollen gathered by bees foraging on fly ash-grown plants. Bull. environ. Contamin. Toxicol., 18: 442-444.

DE WITT, W.B. & SCHWARZ, K. (1958) Multiple dietary necrotic degeneration in the mouse. Experientia, 14: 28-34.

DICKSON, J.D. (1969) Notes on hair and nail loss after ingesting Sapucaia nuts (Lecythis elliptica). Econ. Bot., 23: 133-134.

DICKSON, R.C. & TOMLINSON, R.H. (1967) Selenium in blood and human tissues. Clin. Chim. Acta, 16: 311-321.

DINKEL, C.A., MINYARD, J.A., WHITEHEAD, E.I., & OLSON, O.E. (1957) Agricultural research at the Reed Ranch substation,

Brookings, South Dakota, South Dakota State College of Agriculture and Mechanic Arts, 35 pp (Agricultural Experiment Station Circular No. 135).

DINKEL, C.A., MINYARD, J.A., & RAY, D.E. (1963) Effects of season of breeding on reproductive and weaning performance of beef cattle grazing seleniferous range. J. anim. Sci., 22: 1043-1045.

DIPLOCK, A.T. (1974a) A possible role for trace amounts of selenium and vitamin E in the electron-transfer system of rat-liver microsomes. In: Hoekstra, W.G., Suttie, J.W., Ganther, H.E., & Mertz, W., ed. Trace element metabolism in animals. II, Baltimore, Maryland, University Park Press, pp. 147-160.

DIPLOCK, A.T. (1974b) Possible stabilizing effect of vitamin E on microsomal, membrane-bound, selenide-containing proteins and drug-metabolizing enzyme systems. Am. J. clin. Nutr., 27: 995-1004.

DIPLOCK, A.T. (1976) Metabolic aspects of selenium action and toxicity. CRC Crit. Rev. Toxicol., 4: 271-329.

DIPLOCK, A.T. (1979) The influence of selenium and vitamin E on oxidative demethylation reactions. In: Olive, G., ed. Adv. Pharmacol. Therap., Proceedings of the 7th International Congress on Pharmacology, Vol. 8, pp. 25-33.

DIPLOCK, A.T., BAUM, H., & LUCY, J.A. (1971) The effect of vitamin E on the oxidation state of selenium in rat liver. Biochem. J., 123: 721-729.

DIPLOCK, A.T., CAYGILL, C.P.J., JEFFERY, E.H., & THOMAS, C. (1973) The nature of the acid-volatile selenium in the liver of the male rat. Biochem. J., 134: 283-293.

DONALDSON, W.E. (1977) Selenium inhibition of avian fatty acid synthetase complex. Chem.-biol. Interac., 17: 313-320.

DONGHERTY, J.J. & HOEKSTRA, W.G. (1982) Stimulation of lipid peroxidation in vivo by injected selenite and lack of stimulation by selenate. Proc. Soc. Exp. Biol. Med., 169(2): 209-215.

DORAN, J.W. (1976) Microbial transformations of selenium in soil and culture, Ithaca, New York, Cornell University (Ph.D. Thesis).

DOUGLASS, J.S., MORRIS, V.C., SOARES, J.H., Jr, & LEVANDER, O.A. (1981) Nutritional availability to rats of selenium in tuna, beef kidney, and wheat. J. Nutr., 111: 2180-2187.

DUCE, R.A., HOFFMAN, G.L., & ZOLLER, W.H. (1975) Atmospheric trace metals at remote northern and southern hemisphere sites: pollution or natural? Science, 187: 59-61.

DUDLEY, H.C. (1938) Toxicology of selenium. V. Toxic and vesicant properties of selenium oxychloride. Public Health Rep., 53: 94-98.

DUDLEY, H.C. & MILLER, J.W. (1941) Toxicology of selenium. VI. Effects of subacute exposure to hydrogen selenide. J. ind. Hyg. Toxicol., 23: 470-477.

DUTKIEWICZ, T., DUTKIEWICZ, B., & BALCERSKA, I. (1972) Dynamics of organ and tissue distribution of selenium after intragastric and dermal administration of sodium selenite. Bromatol. Chem. Toksykol., 4: 475-481.

DUVOIR, M., POLLETT, L., & HERRENSCHMIDT, J.L. (1937) Eczéma professionnel dû au sélénium. Bull. Soc. Franc. Dermat. Syphil., 44: 88-95.

DYER, D.G., SCHUETT, V.E., & GANTHER, H.E. (1977) Blood selenium and glutathione peroxidase in children with PKU on diet therapy. In: Abstracts of the Western Hemisphere Nutrition Congress. V, Quebec, pp. 417.

EGAN, A., KERR, S., & MINSKI, M.J. (1977) Instrumental neutron activation analysis of selenium using $^{77}\mu Se$ (T 1/2 = 17s) in biological materials. In: Brown, S.S., ed. Clinical chemistry and chemical toxicology of metals, Amsterdam, Elsevier North-Holland Biomedical Press, pp. 353-356.

ELLIS, N., LLOYD, B., LLOYD, R.S., & CLAYTON, B.E. (1984) Selenium and vitamin E in relation to risk factors for coronary heart disease. J. Clin. Pathol, 37(2): 200-206.

ENGBERG, R.A. (1973) Selenium in Nebraska's groundwater and streams, Lincoln, Nebraska, University of Nebraska (Nebraska Water Survey Paper No. 35).

ERMAKOV, V.V. (1975) [Selenium determination in biological materials by fluorometry and gas-chromatography.] In: [Vitamins. VIII. Biochemistry of vitamin E and selenium,] Kiev, Naukova Dumka Publishing House, pp. 141-146 (in Russian).

ERMAKOV, V.V. & KOVALSKIJ, V.V. (1968) [The geochemical ecology of organisms at high selenium levels in the environment.] In: [Transactions of the biogeochemical laboratory,] Moscow, Nauka Publishing House, Vol. 12, pp. 204-237 (in Russian).

ERMAKOV, V.V. & KOVALSKIJ, V.V. (1974) [The biological importance of selenium,] Moscow, Nauka Publishing House, 298 pp (in Russian).

ERSHOV, V.P. (1969) [Hygienic aspects of selenium containing steel production and the prevention of selenium intoxication.] Gig. Tr. Prof. Zabol., 12: 29-33 (in Russian).

EVANS, C.S., ASHER, C.J., & JOHNSON, C.M. (1968) Isolation of dimethyldiselenide and other volatile selenium compounds from Astragalus racemosus (Pursh.). Austral. J. biol. Sci., 21: 13-20.

EVANS, H.M. & BISHOP, K.S. (1922) On the existence of a hitherto unrecognized dietary factor essential for reproduction. Science, 56: 650-651.

EWAN, R.C., POPE, A.L., & BAUMANN, C.A. (1967) Elimination of fixed selenium by the rat. J. Nutr., 91: 547-554.

EWAN, R.C., BAUMANN, C.A., & POPE, A.L. (1968) Retention of selenium by growing lambs. J. agric. food Chem., 16: 216-219.

EXON, J.H., KOLLER, L.D., & ELLIOTT, S.C. (1976) Effect of dietary selenium on tumor induction by an oncogenic virus. Clin. Toxicol., 9: 273-279.

FALK, R. & LINDHE, J.C. (1974) Radiation dose received by humans from intravenously-administered sodium selenite marked with selenium-75, Los Alamos, New Mexico, Los Alamos Scientific Laboratory (Los Alamos Translation LA-TR-76-5).

FENG, Z., WANG, X., & HAN, C. (1985) [Studies on the toxicity of diets with different selenium levels in rats.] Food Hyg. Res., 3(2): 14-20 (in Chinese).

FERRETTI, R.J. & LEVANDER, O.A. (1974) Effect of milling and processing on the selenium content of grains and cereal products. J. agric. food Chem., 22: 1049-1051.

FERRETTI, R.J. & LEVANDER, O.A. (1976) Selenium content of soybean foods. J. agric. food Chem., 24: 54-56.

FILATOVA, V.S. (1948) [Characteristics of selenium as an industrial intoxicant.] Can. Med. Thesis (in Russian).

FILATOVA, V.S. (1951) [On the toxicity of selenium chloride.] Gig. i Sanit., 5: 18-23 (in Russian).

FLEMING, G.A. (1962) Selenium in Irish soils and plants. Soil Sci., 94: 28-35.

FLEMING, G.A. & WALSH, T. (1957) Selenium occurrence in certain Irish soils and its toxic effects on animals. Proc. Royal Irish Acad., 58: 151-166.

FLOHE, L., GUNZLER, W.A., & SCHOEKS, H.H. (1973) Glutathione peroxidase: a selenoenzyme. FEBS Lett., 32: 132-134.

FORBES, S. & BOUND, G.P. (1977) Some investigations into the electroanalytical chemistry of selenium. Proc. Anal. Div. Chem. Soc., 14: 253-256.

FORSTROM, J.W., ZAKOWSKI, J.J., & TAPPEL, A.L. (1978) Identification of the catalytic site of rat liver glutathione peroxidase as selenocysteine. Biochemistry, 17: 2639-2644.

FRANKE, K.W. & MOXON, A.L. (1936) A comparison of the minimum fatal doses of selenium, tellurium, arsenic, and vanadium. J. Pharmacol. exp. Ther., 58: 454-459.

FRANKE, K.W. & POTTER, V.R. (1936) The effect of selenium containing foodstuffs on growth and reproduction of rats at various ages. J. Nutr., 12: 205-214.

FRANKE, K.W. & TULLY, W.C. (1935) A new toxicant occurring naturally in certain samples of plant foodstuffs. V. Low hatchability due to deformities in chicks. Poultry Sci., 14: 273-279.

FRANKE, K.W., MOXON, A.L., POLEY, W.E., & TULLY, W.C. (1936) Monstrosities produced by the injection of selenium salts into hen's eggs. Anat. Rec., 65: 15-22.

FROSETH, J.A., PIPER, R.C., & CARLSON, J.R. (1974) Relationship of dietary selenium and oral methyl mercury to blood and tissue selenium and mercury concentrations and deficiency: toxicity signs in swine. Fed. Proc., 33: 660.

FROST, D.V. & INGVOLDSTAD, D. (1975) Ecological aspects of selenium and tellurium in human and animal health. Chemica Scripta, 8A: 96-107.

FURR, A.K., PARKINSON, T.F., HINRICHS, R.A., VAN CAMPEN, D.R., BACHE, C.A., GUTENMANN, W.H., ST. JOHN, L.E., Jr, PAKKALA, I.S., & LISK, D.J. (1977) National survey of elements and radioactivity in fly ashes. Absorption of elements by cabbage grown in fly ash-soil mixtures. Environ. Sci. Technol., 11: 1194-1201.

FURR, A.K., PARKINSON, T.F., GUTENMANN, W.H., PAKKALA, I.S., & LISK, D.J. (1978a) Elemental content of vegetables, grains, and forages field-grown on fly ash amended soil. J. agric. food Chem., 26: 357-359.

FURR, A.K., PARKINSON, T.F., HEFFRON, C.L., REID, J.T., HASCHEK, W.M., GUTENMANN, W.H., PAKKALA, I.S., & LISK, D.J. (1978b) Elemental content of tissues of sheep fed rations containing coal fly ash. J. agric. food Chem., 26: 1271-1274.

GAIROLA, C. & CHOW, C.K. (1982) Dietary selenium, hepatic arylhydrocarbon hydroxylase, and mutagenic activation of benzo(a)pyrene, 2-aminoanthracene, and 2-aminofluorene. Toxicol. Lett., 11: 281-287.

GANAPATHY, S. & DHANDA, R. (1976) Selenium content of omniverous and vegetarian diets. Fed. Proc., 35: 360.

GANAPATHY, S.N., JOYNER, B.T., & HAFNER, K.M. (1975) Effect of baking, broiling, and frying on the selenium content of selected beef, pork, chicken and fish foods. In: Proceedings of the 10th International Congress of Nutrition - Abstracts, Kyoto, Japan, 3-9 August, 1975, p. 277.

GANTHER, H.E. (1968) Selenotrisulfides. Formation by the reaction of thiols with selenious acid. Biochemistry, 7: 2898-2905.

GANTHER, H.E. (1971) Reduction of the selenotrisulfide derivative of glutathione to a persulfide analog by glutathione reductase. Biochemistry, 10: 4089-4098.

GANTHER, H.E. (1978) Modification of methylmercury toxicity and metabolism by selenium and vitamin E: possible mechanisms. Environ. Health Perspect., 25: 71-76.

GANTHER, H.E. (1979) Metabolism of hydrogen selenide and methylated selenides. Adv. Nutr. Res., 2: 107-128.

GANTHER, H.E. & BAUMANN, C.A. (1962) Selenium metabolism. II. Modifying effects of sulfate. J. Nutr., 77: 408-414.

GANTHER, H.E. & CORCORAN, C. (1969) Selenotrisulfides. II. Cross-linking of reduced pancreatic ribonuclease with selenium. Biochemistry, 8: 2557-2563.

GANTHER, H.E. & HSIEH, H.S. (1974) Mechanisms for the conversion of selenite to selenides in mammalian tissues. In: Hoekstra, W.G., Suttie, J.W., Ganther, H.E., & Mertz, W., ed. Trace element metabolism in animals. 2, Baltimore, Maryland, University Park Press, pp. 339-353.

GANTHER, H.E. & SUNDE, M.L. (1974) Effect of tuna fish and selenium on the toxicity of methylmercury: a progress report. J. food Sci., 39: 1-5.

GANTHER, H.E., LEVANDER, O.A., & BAUMANN, C.A. (1966) Dietary control of selenium volatilization in the rat. J. Nutr., 88: 55-60.

GANTHER, H.E., GOUDIE, C., SUNDE, M.L., KOPECKY, M.J., WAGNER, P., OH, S., & HOEKSTRA, W.G. (1972) Selenium: relation to decreased toxicity of methylmercury added to diets containing tuna. Science, 175: 1122-1124.

GANTHER, H.E., HAFEMAN, D.G., LAWRENCE, R.A., SERFASS, R.E., & HOEKSTRA, W.G. (1976) Selenium and glutathione peroxidase in health and disease: a review. In: Prasad, H.H., ed. Trace elements in human health and disease. II. Essential and toxic elements, New York, Academic Press pp. 165-234.

GARDINER, M.R., ARMSTRONG, J., FELS, H., & GLENCROSS, R.N. (1962) A preliminary report on selenium and animal health in Western Australia. Aust. J. exp. Agric. Animal Husb., 2: 261-269.

GARDNER, S. (1973) Selenium in animal feed. Proposed food additive regulation. Fed. Reg., 38: 10458-10460.

GE, K., XUE, A., BAI, J., & WANG, S. (1983) Keshan disease - an endemic cardiomyopathy in China. Virchows Arch. (Pathol. Anat.), 401: 1-15.

GEERING, H.R., CARY, E.E., JONES, L.H.P., & ALLAWAY, W.H. (1968) Solubility and redox criteria for the possible forms of selenium in soils. Soil Sci. Soc. Am. Proc., 32: 35-40.

GIASUDDIN, A.S.M., CAYGILL, C.P.J., DIPLOCK, A.T., & JEFFERY, E.H. (1975) The dependence on vitamin E and selenium of drug demethylation in rat liver microsomal fractions. Biochem. J., 146: 339-350.

GISSEL-NIELSEN, G. (1971) Selenium content of some fertilizers and their influence on uptake of selenium in plants. J. agric. food Chem., 19: 564-566.

GISSEL-NIELSEN, G. (1973) Ecological effects of selenium application to field crops. Ambio, 2: 114-117.

GISSEL-NIELSEN, G. (1976) Selenium in soils and plants. In: Proceedings of the Symposium on Se-Te in the Environment, Pittsburgh, Pennsylvania, Industrial Health Foundation, pp. 10-25.

GISSEL-NIELSEN, G. (1986) Selenium fertilizers and foliar application, Danish experiments. Ann. clin. Res., 18: 61-64.

GISSEL-NIELSEN, M. & GISSEL-NIELSEN, G. (1975) Selenium in soil-animal relationships. Pedobiologia, 15: 65-67.

GITSOVA, S. (1973) Presence of selenium in drinking waters in Bulgaria. Khig. Zdraveopazvane, 16: 557-561.

GLOVER, J.R. (1954) Some medical problems concerning selenium in industry. Trans. Assoc. Ind. Med. Off., 4: 94-96.

GLOVER, J.R. (1967) Selenium in human urine: a tentative maximum allowable concentration for industrial and rural populations. Ann. occup. Hyg., 10: 3-14.

GLOVER, J.R. (1970) Selenium and its industrial toxicology. Ind. Med. Surg., 39: 50-54.

GLOVER, J.R. (1976) Environmental health aspects of selenium and tellurium. In: Proceedings of the Symposium on Selenium-Tellurium in the Environment, Pittsburgh, Pennsylvania, Industrial Health Foundation, pp. 279-292.

GODWIN, K.O. & FUSS, C.N. (1972) The entry of selenium into rabbit protein following the administration of Na$_2$ ^{75}SeO$_3$. Aust. J. biol. Sci., 25: 865-871.

GOODWIN, W.J., LANE, H.W., BRADFORD, K., MARSHALL, M.V., GRIFFIN, A.C., GEOPFERT, H., & JESSE, R.H. (1983) Selenium and glutathione peroxidase levels in patients with epidermoid carcinoma of the oral cavity and orophasynx. Cancer, 51: 110-115.

GORTNER, R.A., Jr (1940) Chronic selenium poisoning of rats as influenced by dietary protein. J. Nutr., 19: 105-112.

GOULDEN, P.D. & BROOKSBANK, P. (1974) Automated atomic absorption determination of arsenic, antimony, and selenium in natural waters. Anal. Chem., 46: 1431-1436.

GRACIANSKAJA, L.N. & KOVSHILO, V.E., ed. (1977) [Selenium and its compounds]. In: [Handbook of occupational pathology,] Moscow, Medicina Publishing House pp. 330-332 (in Russian).

GRANT, K.E., CONNER, M.W., & NEWBERNE, P.M. (1977) Effect of dietary sodium selenite upon lesions induced by repeated small doses of aflatoxin B_1. Toxicol. appl. Pharmacol., 41: 166.

GREEN, J., BUNYAN, J., CAWTHORNE, M.A., & DIPLOCK, A.T. (1969) Vitamin E and hepatotoxic agents. I. Carbon tetrachloride and lipid peroxidation in the rat. Br. J. Nutr., 23: 297-307.

GRIFFIN, A.C. & JACOBS, M.M. (1977) Effects of selenium on azo dye hepatocarcinogenesis. Cancer Lett., 3: 177-181.

GRIFFITHS, N.M. (1973) Dietary intake and urinary excretion of selenium in some New Zealand women. Proc. Univ. Otago Med. School, 51: 8-9.

GRIFFITHS, N.M. & THOMSON, C.D. (1974) Selenium in the whole blood of New Zealand residents. N.Z. med. J., 80: 199-202.

GRIFFITHS, N.M., STEWART, R.D.H., & ROBINSON, M.F. (1976) The metabolism of ^{75}Se-selenomethionine in four women. Br. J. Nutr., 35: 373-382.

GRIMANIS, A.P., VASSILAKI-GRIMANI, M., ALEXIOU, D., & PAPADATOS, C. (1978) Determination of seven trace elements in human milk, powdered cow's milk, and infant foods by neutron activation analysis. In: Nuclear activation techniques in the life sciences, Vienna, International Atomic Energy Agency, pp. 241-253.

GROCE, A.W., MILLER, E.R., KEAHEY, K.K., ULLREY, D.E., & ELLIS, D. J. (1971) Selenium supplementation of practical diets for growing-finishing swine. J. Ann. Sci., 32: 905-911.

GROSS, S. (1976) Hemolytic anemia in premature infants: Relationship to vitamin E, selenium, glutathione peroxidase, and erythrocyte lipids. Semin. Hematol., 13: 187-199.

GROSSMAN, A. & WENDEL, A. (1983) Non-reactivity of the selenoenzyme glutathione peroxidase with enzymatically hydroperoxidized phospholipids. Eur. J. Biochem., 135(3): 549-552.

GU, B. (1983) Pathology of Keshan disease: a comprehensive review. Chin. med. J., 96: 251-261.

GUSEJNOV, T.M., ZHAFAROV, A.I., PERELYGIN, V.V., & KARAEV, M.A. (1974) [On the localization of endogenous selenium in structural slements of bull's eye.] In: [Selenium in biology,] Baku, Elm Publishing House, pp. 47-49.

GUTENMANN, W.H., BACHE, C.A., YOUNGS, W.D., & LISK, D.J. (1976) Selenium in fly ash. Science, 191: 966-967.

HADDAD, P.R. & SMYTHE, L.E. (1974) A critical evaluation of fluorometric methods for determination of selenium in plant materials with 2,3-diaminoaphthalene. Talanta, 21: 859-865.

HADJIMARKOS, D.M. (1956) Geographic variations of dental caries in Oregon. VII. Caries prevalence among children in the blue mountains region. J. Pediatr., 48: 195-201.

HADJIMARKOS, D.M. (1960) Urinary selenium and dental caries. Nature (Lond.), 188: 677.

HADJIMARKOS, D.M. & BONHORST, C.W. (1958) The trace element selenium and its influence on dental caries susceptibility. J. Pediatr., 52: 274-278.

HADJIMARKOS, D.M. & BONHORST, C.W. (1961) The selenium content of eggs, milk, and water in relation to dental caries in children. J. Pediatr., 59: 256-259.

HADJIMARKOS, D.M. & SHEARER, T.R. (1971) Selenium concentration in human saliva. Am. J. clin. Nutr., 24: 1210.

HADJIMARKOS, D.M. & STORVICK, C.A. (1950) Geographic variations of dental caries in Oregon. IV. Am. J. public Health, 40: 1552-1555.

HADJIMARKOS, D.M., STORVICK, C.A., & REMMERT, L.F. (1952) Selenium and dental caries. An investigation among school children of Oregon. J. Pediatr., 40: 451-455.

HAFEMAN, D.G. & HOEKSTRA, W.G. (1977a) Protection against carbon tetrachloride-induced lipid peroxidation in the rat by dietary vitamin E, selenium, and methionine as measured by ethane evolution. J. Nutr., 107: 656-665.

HAFEMAN, D.G. & HOEKSTRA, W.G. (1977b) Lipid peroxidation in vivo during vitamin E and selenium deficiency in the rat as monitored by ethane evolution. J. Nutr., 107: 666-672.

HAFEMAN, D.G., SUNDE, R.A., & HOEKSTRA, W.G. (1974) Effect of dietary selenium on erythrocyte and liver glutathione peroxidase in the rat. J. Nutr., 104: 580-587.

HAKKARAINEN, J., LINDBERG, P., BENGTSSON, G., JONSSON, L., & LANNEK, N. (1978) Requirement for selenium (as selenite) and vitamin E (as α-tocopherol) in weaned pigs. III. The effect on the development of the VESD syndrome of varying selenium levels in a low-tocopherol diet. J. anal. Sci., 46: 1001-1008.

HALL, R.H., LASKIN, S., FRANK, P., MAYNARD, E.A., & HODGE, C.H. (1951) Arch. ind. Hyg. occup. Med., 4: 458-464.

HALTER, K. (1938) [Selenium poisoning, especially skin changes accompanied by secondary porphysia.] Arch. Dermatol., 178: 340 (in German).

HALVERSON, A.W. (1974) Growth and reproduction with rats fed selenite-Se. Proc. S. Dak. Acad. Sci., 53: 167-177.

HALVERSON, A.W., HENDRICK, C.M., & OLSON, O.E. (1955) Observations on the protective effect of linseed oil meal and some extracts against chronic selenium poisoning in rats. J. Nutr., 56: 51-60.

HALVERSON, A.W., GUSS, P.L., & OLSON, O.E. (1962) Effect of sulfur salts on selenium poisoning in the rat. J. Nutr., 77: 459-464.

HALVERSON, A.W., PALMER, I.S., & GUSS, P.L. (1966) Toxicity of selenium to post-weanling rats. Toxicol. appl. Pharmacol., 9: 477-484.

HAMDY, A.A. & GISSEL-NIELSEN, G. (1976) Volatilization of selenium from soils. Z. Pflanzenernaehr. Bodenkd., 6: 671-678.

HAMILTON, A. (1927) Industrial poisons in the United States, Macmillan Company, p. III.

HAMILTON, A. (1934) Industrial toxicology, New York, Harpers Publishing Company, p. 111.

HAMILTON, J.W. (1975) Chemical examination of seleniferous cabbage Brassica oleracea capitata. J. agric. food Chem., 23: 1150-1152.

HARLAND, B.F., PROSKY, L., & VANDERVEEN, J.E. (1978) Nutritional adequacy of current levels of Ca, Cu, Fe, I, Mg, Mn, P, Se, and Zn in the American food supply for adults, infants, and toddlers. In: Kirchgessner, M., ed. Trace element metabolism in man and animals. 3, Freising- Weihenstephan, West Germany, Arbeitskreis fur Tierernährungsforschung Weihenstephan, pp. 311-315.

HARR, J.R. & MUTH, O.H. (1972) Selenium poisoning in domestic animals and its relationship to man. Clin. Toxicol., 5: 175-186.

HARR, J.R., BONE, J.F., TINSLEY, I.J., WESWIG, P.H., & YAMAMOTO, R.S. (1967) Selenium toxicity in rats. II. Histopathology. In: Muth, O.H., Oldfield, J.E., & Weswig, P.H., ed. Selenium in biomedicine, Westport, Connecticut, The AVI Publishing Co., Inc, pp. 153-178.

HARR, J.R., EXON, J.H., WHANGER, P.D., & WESWIG, P.H. (1972) Effect of dietary selenium on N-2-fluorenyl-acetamide (FAA)-induced cancer in vitamin E-supplemented, selenium depleted rats. Clin. Toxicol., 5: 187-194.

HARRIS, P.L., LUDWIG, M.I., & SCHWARZ, K. (1958) Ineffectiveness of Factor 3-active selenium compounds in resorption-gestation bioassay for vitamin E. Proc. Soc. Exp. Biol. Med., 97: 686-688.

HARTHOORN, A.M. & YOUNG, E. (1974) A relationship between acid-base balance and capture myopathy in zebra (Equus burchelli) and an apparent therapy. Vet. Res., 95: 337-342.

HARTLEY, W.J. (1963) Selenium and ewe fertility. Proc. N.Z. Soc. Anim. Prod., 23: 20-27.

HARTLEY, W.J. (1967) Levels of selenium in animal tissues and methods of selenium administration . In: Muth, O.H., Oldfield, J.E., & Weswig, P.H., ed. Selenium in biomedicine, Westport, Connecticut, The AVI Publishing Co., Inc, pp. 79-96.

HARTLEY, W.J. & GRANT, A.B. (1961) A review of selenium responsive diseases of New Zealand livestock. Fed. Proc., 20: 679-688.

HARTLEY, W.J., GRANT, A.B., & DRAKE, C. (1960) Control of white muscle disease and ill thrift with selenium. N.Z. J. Agric., 101: 343-345.

HASHIMOTO, Y. & WINCHESTER, J.W. (1967) Selenium in the atmosphere. Environ. Sci. Technol., 1: 338-340.

HEATH, R.L. (1969-70) Handbook of chemistry and physics, 50th ed., Cleveland, Ohio, The Chemical Rubber Publishing Co.

HE, G. (1979) On the etiology of Keshan disease: two hypotheses. Chin. med. J., 92: 416-422.

HEINRICH, M., Jr & KELSEY, F.E. (1955) Studies on selenium metabolism: the distribution of selenium in the tissues of the mouse. J. Pharmacol. exp. Therap., 114: 28-32.

HELZLSOUER, K., JACOBS, R., & MORRIS, S. (1985) Acute selenium intoxication in the United States. Fed. Proc., 44(5): 1670.

HICKEY, F. (1968) Selenium in human and animal nutrition. N.Z. Agric., 18: 1-2.

HICKEY, F. (1977) Human medication with selenium. Including brief comments on the comparative pathology of White Muscle Disease in livestock and human muscular dystrophies. In: Hamilton, N.Z., ed. Trace elements in human and animal health in New Zealand, Hamilton, New Zealand, Waikato University Press, pp. 92-99.

HIGGS, D.J., MORRIS, V.C., & LEVANDER, O.A. (1972) Effect of cooking on selenium content of foods. J. agric. food Chem., 20: 678-680.

HODDER, A.P.W. & WATKINSON, J.H. (1976) Low selenium levels in tephra-derived soil - inherent or pedogenetic? N.Z. J. Sci., 19: 397-400.

HOEKSTRA, W.G. (1975a) Biochemical function of selenium and its relation to vitamin E. Fed. Proc., 34: 2083-2089.

HOEKSTRA, W.G. (1975b) Glutathione peroxidase activity of animal tissues as an index of selenium status. In: Hemphill, D.D., ed. Trace substances in environmental health. IX, Columbia, Missouri, University of Missouri Press, pp. 331-337.

HOEKSTRA, W.G., HAFEMAN, O., OH, S.H., SUNDER, R.A., & GAN-THER, H.E. (1973) Effect of dietary selenium on liver and erythrocyte glutathione peroxidase in the rat. Fed. Proc., 32: 885.

HOGGER, D. & BOHM, C. (1944) [On dermal injury produced by selenite]. Dermatologica, 90: 217-223 (in German).

HOLMBERG, R.E., & FERM, V.H. (1969) Interrelationships of selenium, cadmium, and arsenic in mammalian teratogenesis. Arch. environ. Health, 18: 873-877.

HOLSTEIN, E. (1951) [The effects of occupational exposure to selenium.] Zentrblt. Arbeitsmed. Arbeitschutz, 1: 102-104 (in German).

HOLYNSKA, B. & MARKOWICZ, A. (1977) Application of energy dispersive X-ray fluorescence for the determination of selenium in blood and tissue. Radiochem. radioanal. Lett., 31: 165-170.

HOPKINS, L.L. & MAJAJ, A.S. (1967) Selenium in human nutrition. In: Muth, O.H., Oldfield, J.E., & Weswig, P.H., ed. Selenium in biomedicine, Westport, Connecticut, The AVI Publishing Co., Inc, pp. 203-214.

HOVE, E.L. (1948) Interrelation between α-tocopherol and protein metabolism. III. The protective effect of vitamin E and certain nitrogenous compounds against CCl_4 poisoning in rats. Arch. Biochem., 17: 467-474.

HOWARD, J.H., III (1971) Control of geochemical behavior of selenium in natural waters by adsorption on hydrous ferric oxides. In: Hemphill, D.D., ed. Trace substances in environmental health. V, Columbia, Missouri, University of Missouri Press, pp. 485-495.

HOWE, M. (1974) Selenium in the blood of south Dakotans. Arch. environ. Health, 34: 444-448.

HURT, H.D., CARY E.E., & VISEK, W.J. (1971) Growth, reproduction, and tissue concentrations of selenium in the selenium-depleted rat. J. Nutr., 101: 761-766.

IARC (1975) Some aziridines, N,-S-, and O-mustards and selenium, Lyons, International Agency for Research on Cancer, pp. 245-260 (IARC Monographs on the Evaluation of the Carcinogenic Risk of Chemicals to Man, Vol. 9).

IHNAT, M. (1974) Collaborative study of a fluorometric method for determining selenium in foods. J. Assoc. Off. Anal. Chem., 57: 373-378.

IHNAT, M. (1976) Selenium in foods: evaluation of atomic absorption spectrometric techniques involving hydrogen selenide generation and carbon furnace atomization. J. Assoc. Off. Anal. Chem., 59: 911-922.

IHNAT, M. & MILLER, H.J. (1977a) Analysis of foods for arsenic and selenium by acid digestion, hydride evolution atomic absorption spectrophotometry. J. Assoc. Off. Anal. Chem., 60: 813-825.

IHNAT, M. & MILLER, H.J. (1977b) Acid digestion, hydride evolution atomic absorption spectrophotometric method for determining arsenic and selenium in foods: collaborative study. I. J. Assoc. Off. Anal. Chem., 60: 1414-1433.

INNES, J.R.M., ULLAND, B.M., VALERIO, M.G., PETRUCELLI, L., FISHBEIN, L., HART, E.R., PALLOTTA, A.J., BATES, R.R., FALK, H.L., GART, J.J., KLEIN, M., MITCHELL, I., & PETERS, J. (1969) Bioassay of pesticides and industrial chemicals for tumorigenicity in mice: a preliminary note. J. Natl Cancer Inst., 42: 1101-1114.

IP, C. (1985a) Selenium inhibition of chemical carcinogenesis. Fed. Proc., 44: 2573-2578.

IP, C. (1985b) Attenuation of the anticarcinogenic action of selenium by vitamin E deficiency. Cancer Lett., 25: 325-331.

IP, C. & SINHA, D.K. (1981) Enhancement of mammary tumorigenesis by dietary selenium deficiency in rats with a high polyunsaturated fat intake. Cancer Res., 41: 31-34.

IRGOLIC, K.J. & KUDCHADKER, M.H. (1974) Organic chemistry of selenium. In: Zingaro, R.A. & Cooper, W.C., ed. Selenium, New York, Van Nostrand Reinhold Co., pp. 408-545.

IZRAELSON, Z.I., MOGILEVSKAJA, O.Ja., & SUVOROV, S.W. (1973) [Selenium.] In: [Problems of occupational hygiene and occupational pathology connected with the work with toxic metals,] Moscow, Medicina Publishing House pp. 245-257 (in Russian).

JACOBS, M.M. (1976) Selenium: a possible inhibitor of colon and rectum cancer. II. Biochemical aspects. In: Proceedings of the Symposium on Selenium-Tellurium in the Environment, Pittsburgh, Pennsylvania, Industrial Health Foundation, pp. 329-340.

JACOBS, M.M. (1977) Inhibitory effects of selenium on 1,2-dimethylhydrazine and methylazoxymethanol colon carcinogenesis. Cancer, 40: 2557-2564.

JACOBS, M.M. & FORST, C. (1981a) Toxicological effects of sodium selenite in 5 prague-Dawley rats. J. Toxicol. environ. Health, 8: 575-585.

JACOBS, M.M. & FORST, C. (1981b) Toxicological effects of sodium selenite in Swiss mice. J. Toxicol. environ. Health, 8: 587-598.

JACOBS, M.M., JANSSON, B., & GRIFFIN, A.C. (1977a) Inhibitory effects of selenium on 1,2-dimethylhydrazine and methylazoxymethanol acetate induction of colon tumours. Cancer Lett., 2: 133-138.

JACOBS, M.M., MATNEY, T.S., & GRIFFIN, A.C. (1977b) Inhibitory effects of selenium on the mutagenicity of 2-acetylaminofluorene (AAF) and AAF derivatives. Cancer Lett., 2: 319-322.

JACOBSSON, S.O., LIDMAN, S., & LINDBERG, P. (1970) Blood selenium in a beef herd affected with muscular degeneration. Acta vet. Scand., 11: 324-326.

JAFFE, W.G. (1973) Selenium in food plants and feeds. Toxicology and nutrition. Qualitas Plantarum. Plant foods hum. Nutr., 23: 191-204.

JAFFE, W.G. (1976) Effect of selenium intake in humans and in rats. In: Proceedings of the Symposium on Selenium-Tellurium in the Environment, Pittsburgh, Pennsylvania, Industrial Health Foundation, pp. 188-193.

JAFFE, W.G. & MONDRAGON, M.C. (1969) Adaptation of rats to selenium intake. J. Nutr., 97: 431-436 .

JAFFE, W.G. & MONDRAGON, C. (1975) Effects of ingestion of organic selenium in adapted and non-adapted rats. Br. J. Nutr., 33: 387-397.

JAFFE, W.G. & VELEZ, B.F. (1973) Selenium intake and congenital malformations in humans. Arch. Latinoam. Nutr., 23: 514-516.

JAFFE, W.G., RUPHAEL, M.D., MONDRAGON, M.C., & CUEVAS, M.A. (1972a) Clinical and biochemical studies on school children from a seleniferous zone. Arch. Latinoam. Nutr., 22: 595-611.

JAFFE, W.G., MONDRAGON, M.C., LAYRISSE, M., & OJEDA, A. (1972b) Toxicity symptoms in rats fed organic selenium. Arch. Latinoam. Nutr., 22: 467-481.

JANSSON, B., JACOBS, M.M., & GRIFFIN, A.C. (1978) Gastrointestinal cancer: epidemiology and experimental studies. In: Schrauzer, G.N., ed. Inorganic and nutritional aspects of cancer, New York, Plenum Press, pp. 305-322.

JENKINS, K.J. & HIDIROGLOU, M. (1971) Comparative uptake of selenium by low cystine and high cystine proteins. Can. J. Biochem., 49: 468-472.

JENSEN, L.S. & MCGINNIS, J. (1960) Influence of selenium, antioxidants, and type of yeast on vitamin E deficiency in the adult chicken. J. Nutr., 72: 23-28.

JENSEN, R., CLOSSON, W., & ROTHENBERG, R. (1984) Selenium intoxication - New York. Mobid. Mortal. Weekly Rep., 33: 157-158.

JOHNSON, C.M. (1976) Selenium in the environment. Res. Rev., 62: 101-130.

JONES, G.B. & GODWIN, K.O. (1962) Distribution of radioactive selenium in mice. Nature (Lond.), 196: 1294-1296.

JUDSON, G.J. & OBST, J.M. (1975) Diagnosis and treatment of selenium inadequacies in the grazing ruminant. In: Nicholas, D.J.D. & Egan, A.R., ed. Trace elements in soil-plant-animal systems, New York, Academic Press, Inc, pp. 385-405.

JULIUS, A.D., DAVIES, M.H., & BIRT, D.F. (1980) Relationship between blood selenium and erythrocyte glutathione peroxidase activity with excess dietary selenium in Syrian golden hamsters. Fed. Proc., 39: 555.

KAMADA, T., SHIRAISHI, T., & YAMAMOTO, Y. (1978) Differential determination of selenium (IV) and selenium (VI) with sodium diethyldithiocarbamate, ammonium pyrrolidinedithiocarbamate and dithizone by atomic-absorption spectrophotometry with a carbon-tube atomizer. Talanta, 25: 15-19.

KAQUELER, J.C., MALOIGNE, E., & BONIFAY, P. (1977) Effects of sodium selenate on caries incidence on the rat. J. dent. Res., D 56 (Special Issue): D151.

KASIMOV, R.Ju., RUSTAMOVA, Sh.A., GUSEJNOV, T.M., & KIAZIMOV, S.K. (1976) [Dynamics of selenium distribution in some organs of young fish from selected valuable edible species.] In: [Selenium in biology,] Baku, Elm Publishing House, Vol. 2, pp. 143-144 (in Russian).

KERDEL-VEGAS, F. (1966) The depilatory and cytotoxic action of "Coco de Mono" (Lecythis ollaria) and its relationship to chronic seleniosis. Econ. Bot., 20: 187-195.

KESHAN DISEASE RESEARCH GROUP (1979a) Observations on effect of sodium selenite in prevention of Keshan disease. Chinese med. J., 92: 471-476.

KESHAN DISEASE RESEARCH GROUP (1979b) Epidemiologic studies on the etiologic relationship of selenium and Keshan disease. Chinese med. J., 92: 477-482.

KETTERER, B., BEALE, D., & MEYER, D. (1982) The structure and multiple functions of glutathione transferases. Biochem. Soc. Trans., 10(2): 82-84.

KILNESS, A.W. (1973) Selenium and public health. S.D. J. Med., 26: 17-19.

KILNESS, A.W. & HOCHBERG, F.H. (1977) Amyotrophic lateral sclerosis in a high selenium environment. J. Am. Med. Assoc., 237: 2843-2844.

KIMMERLE, G. (1960) [Comparative studies on the inhalation toxicity of sulfur-selenium and tellurium hexafluoride.] Arch. Toxikol., 18: 140-144 (in German).

KINNIGKEIT, G. (1962) [Studies on workers exposed to selenium in a rectifier plant.] Z. Gesamte Hyg. Grenzgeb., 8: 350-362 (in German).

KLAYMAN, D.L. & GUNTHER, W.H.H. (1973) Organic selenium compounds: their chemistry and biology, New York, John Wiley and Sons, Inc, 1188 pp.

KLEIN, A.K. (1943) Report on selenium. J. Assoc. Offic. Anal. Chem., 26: 346-352.

KLUG, H.L., PETERSEN, D.F., & MOXON, A.L. (1949) The toxicity of selenium analogues of cystine and methionine. Proc. S.D. Acad. Sci., 28: 117-120.

KLUG, H.L., MOXON, A.L., PETERSEN, D.F, & POTTER, V.R. (1950) The in vivo inhibition of succinic dehydrogenase by selenium and its release by arsenic. Arch. Biochem. Biophys., 28: 253-259.

KOEMAN, J.H., VAN DE VEN, W.S.M., DE GOEIJ, J.J.M., TJIOE, P.S., & VAN HAAFTEN, J.L. (1975) Mercury and selenium in marine mammals and birds. Sci. total Environ., 3: 279-287.

KOIVISTOINEN, P., ed. (1980) Mineral elements composition of Finnish foods: N, K, Ca, Mg, P, S, Fe, Cu, Mn, Zn, Mo, Co, Ni, Cr F, Se, Si, Rb, Al, B, Br, Hg, As, Cd, Pb, and ash. Acta agric. Scand., Suppl. 2: 17.

KOIVISTOINEN, P. & HUTTUNEN, J.K. (1985) Selenium deficiency in Finnish food and nutrition: research strategy and measures. In: Mills, C.F., Bremner, I., & Chesters, J.K., ed Trace elements in man and animals-TEMA 5, Slough, Commonwealth Agricultural Bureaux, pp. 925-928.

KOIVISTOINEN, P. & HUTTUNEN, J.K. (1986) Selenium in food and nutrition in Finland. An overview on research and action. Ann. clin. Res., 18: 13-17.

KOSTA, L., BYRNE, A.R., & ZELENKO, V. (1975) Correlation between selenium and mercury in man following exposure to inorganic mercury. Nature (Lond.), 254: 238-239.

KOVALSKIJ, V.V. (1974) [Geochemical ecology,] Moscow, Nauka Publishing House, 299 pp (in Russian).

KOVALSKIJ, V.V. (1978) [Geochemical ecology: a basis for a system of biogeochemical regional characterization.] In: [Biochemical regional characterization: a method for studying the ecological structure of the biosphere,] Moscow, Nauka Publishing House, Vol. 15, pp. 3-21 (in Russian).

KOVALSKIJ, V.V. & ERMAKOV, V.V. (1975) [Problems of experimental geochemical ecology of animals under the conditions of selenium regions and some approaches to the study of the biological role of selenium.] In: [Vitamins. VII. Biochemistry of vitamin E and selenium,] Kiev, Nakove Dunka Publishing House, pp. 80-87 (in Russian).

KRAUSKOPF, K.B. (1955) Sedimentary deposits of rare elements. Econ. Geol. Fiftieth Anniversary, (Part I): 411-463.

KRONBORG, O.J. & STEINNES, E. (1975) Simultaneous determination of arsenic and selenium in soil by neutron-activation analysis. Analyst, 100: 835-837.

KU, P.K., ELY, W.T., GROCE, A.W., & ULLREY, D.E. (1972) Natural dietary selenium α-tocopherol and effect on tissue selenium. J. Ann. Sci., 34: 208-211.

KUBOTA, J., ALLAWAY, W.H., CARTER, D.L., CARY, E.E., & LAZAR, V.A. (1967) Selenium in crops in the United States in relation to selenium-responsive diseases of animals. J. agric. food Chem., 15: 448-453.

KUBOTA, J., CARY, E.E., & GISSEL-NIELSEN, G. (1975) Selenium in rainwater of the United States and Denmark. In: Hemphill, D.D., ed. Trace substances in environmental health. IX, Columbia, Missouri, University of Missouri Press, pp. 123-130.

KULIEVA, E.M., PERELIGIN, V.V., & DZHAFAROV, A.I. (1978) [Isolated retina electroretinogram (ERG) under the conditions of induced lipoperoxidation.] Dokl. Akad. Nauk. Azarbayidzh SSR, 34(2): 85-89 (in Russian).

KURLAND, L.T. (1977) Amyotrophic lateral sclerosis and selenium. J. Am. Med. Assoc., 238: 2365-2366.

LAKIN, H.W. & BYERS, H.G. (1941) Selenium occurrence in certain soils in the United States, with a discussion of related topics, sixth report, Washington DC, US Department of Agriculture, 26 pp (USDA Technical Bulletin No. 783).

LAKIN, H.W. & DAVIDSON, D.F. (1967) The relation of the geochemistry of selenium to its occurrence in soils. In: Muth, O.H., Oldfield, J.E., & Weswig, P.H., ed. Selenium in biomedicine, Westport, Connecticut, The AVI Publishing Co., Inc, pp. 27-56.

LATHROP, K.A., JOHNSTON, R.E., BLAU, M., & ROTHSCHILD, E.O. (1972) Radiation dose to humans from ^{75}Se-L-selenomethionine. J. Nucl. Med., 13 (suppl. No. 6): 7-30.

LAUER D.J. (1947) Selenium poisoning. A review, with emphasis on its industrial aspects, Pittsburgh, Pennsylvania, University of Pittsburgh School of Medicine, pp. 1-47 (Thesis).

LAWRENCE, R.A. & BURK, R.F. (1976) Glutathione peroxidase activity in selenium-deficient rat liver. Biochem. Biophys. Res. Commun., 71: 952-958.

LAWRENCE, R.A. & BURK, R.F. (1978) Species, tissue, and subcellular distribution of non Se-dependent glutathione peroxidase activity. J. Nutr., 108: 211-215.

LAWRENCE, R.A., PARKHILL, L.K., & BURK, R.F. (1978) Hepatic cytosolic non selenium-dependent glutathione peroxidase activity: its nature and the effect of selenium deficiency. J. Nutr., 108: 981-987.

LAWSON, T. & BIRT, D.F. (1983) Enhancement of the repair of carcinogen-induced DNA damage in the hamster pancreas by dietary selenium. Chem. Biol. Interact., 45: 95-104.

LAZAREV, N.V. (1977) [Harmful compounds in the industry,] 7th ed., Leningrad, Chimija, Vol. 3, pp. 74-82 (in Russian).

LAZAREV, N.V. & GADASKINA, I.D., ed. (1977) [Selenium and its compounds.] In: [Noxious substances in industry,] Leningrad, Chimija Publishing House, pp. 75-82 (in Russian).

LEEB, J., BAUMGARTNER, W.A., LYONS, K., & LORBER, A. (1977) Uptake, distribution, and excretion of sodium selenite in rheumatoid subjects. In: Biological implications of metals in the environment, Technical Information Center, Energy Research and Development Administration, pp. 536-546.

LEMLEY, R.E. (1940) Selenium poisoning in the human. Lancet, 60: 528-531.

LEMLEY, R.E. (1943) Observations on selenium poisoning in South and North America. Lancet, 63: 257-258.

LEMLEY, R.E. & MERRYMAN, M.P. (1941) Selenium poisoning in the human. Lancet, 61: 435-438.

LEVANDER, O.A. (1976a) Selected aspects of the comparative metabolism and biochemistry of selenium and sulfur. In: Prasad, A.S., ed. Trace elements in human health and disease, Vol. 2, Essential and toxic elements, New York, Academic Press, pp. 135-163.

LEVANDER, O.A. (1976b) Selenium in foods. In: Proceedings of the Symposium on Selenium-Tellurium in the Environment, Pittsburgh, Pennsylvania, Industrial Health Foundation, pp 26-53.

LEVANDER, O.A. (1977) Metabolic interrelationships between arsenic and selenium. Environ. Health Perspect., 19: 159-164.

LEVANDER, O.A. (1979) Lead toxicity and nutritional deficiencies. Environ. Health Perspect., 29: 115-125.

LEVANDER, O.A. (1982) Selenium: biochemical actions, inter-actions, and some human health implications. In: Clinical, biochemical, and nutritional aspects of trace elements, pp. 345–368.

LEVANDER, O.A. (1983) Consideration in the design of selenium bioavailability studies. Fed. Proc., 42: 1721–1725.

LEVANDER, O.A. (1986) Human and animal nutrition. In: Mertz, M., ed. Trace elements, Vol. 2, New York, Academic Press, pp. 209–279.

LEVANDER, O.A. & BAUMANN, C.A. (1966) Selenium metabolism. VI. Effect of arsenic on the excretion of selenium in the bile. Toxicol. appl. Pharmacol., 9: 106–115.

LEVANDER, O.A. & MORRIS, V.C. (1970) Interactions of methionine, vitamin E, and antioxidants in selenium toxicity in the rat. J. Nutr., 100: 1111–1118.

LEVANDER, O.A. & MORRIS (1984) Dietary selenium levels needed to maintain balance in North American adults consuming self-selected diets. Am. J. clin. Nutr., 39: 809–815.

LEVANDER, O.A., YOUNG, M.L., & MEEKS, S.A. (1970) Studies on the binding of selenium by liver homogenates from rats fed diets containing either casein or casein plus linseed oil meal. Toxicol. appl. Pharmacol., 16: 79–87.

LEVANDER, O.A., MORRIS, V.C., & HIGGS, D.J. (1973a) Acceleration of thiol-induced swelling of rat liver mitochondria by selenium. Biochemistry, 12: 4586–4590.

LEVANDER, O.A., MORRIS, V.C., & HIGGS, D.J. (1973b) Selenium as a catalyst for the reduction of cytochrome c by gluta-thione. Biochemistry, 12: 4591–4595.

LEVANDER, O.A., WELSH, S.O., & MORRIS, V.C. (1980) Erythro-cyte deformability as affected by vitamin E deficiency and lead toxicity. Ann. N.Y. Acad. Sci., 355: 227.

LEVANDER, O.A., SUTHERLAND, B., MORRIS, V.C., & KING, J.C. (1981a) Selenium metabolism in human nutrition. In: Spallhoz, J.E., Martin, J.L., & Ganther, H.E., ed. Selenium in biology and medicine, Westport, Connecticut, The AVI Publishing Co., Inc, pp. 256–268.

LEVANDER, O.A., SUTHERLAND, B., MORRIS, V.C., & KING, J.C. (1981b) Selenium balance young men during selenium depletion and repletion. Am. J. clin. Nutr., 34: 2662-2669.

LEVANDER, O.A., ALFTHAN, G., ARVILOMMI, H., GREF, C.G., HUTTUNEN, J.K., KATAJA, M., KOIVISTOINEN, P., & PIKKARAINEN, J. (1983) Bioavailability of selenium to Finnish men as assessed by platelet glutathione peroxidase activity and other blood parameters. Am. J. clin. Nutr., 37: 887-897.

LEVINE, R.J. & OLSON, R.E. (1970) Blood selenium in Thai children with protein-calorie malnutrition. Proc. Soc. Exp. Biol. Med., 134: 1030-1034.

LEWIS, B.G., JOHNSON, C.M., & BROYER, T.C. (1974) Volatile selenium in higher plants: the production of dimethyl selenide in cabbage leaves by enzymatic cleavage of Se-methyl seleno-methionine selenonium salt. Plant Soil, 40: 107-118.

LI, C., HUANG, J., & LI, C. (in press) Observational report on the effects of taking sodium selenite for six years continuously as a preventive for Kashin-Beck disease as shown in X-ray studies. In: Proceedings of the Third International Symposium on Selenium in Biology and Medicine.

LIANG, S. (1985) The prophylactic and curing effect of selenium (Se) in combatting of the Kaschin-Beck's disease.

LIEBSCHER, K. & SMITH, H. (1968) Essential and non-essential trace elements. A method of determining whether an element is essential or nonessential in human tissue. Arch. environ. Health, 17: 881-890.

LINDBERG, P. (1968) Selenium determination in plant and animal material, and in water. A methodological study. Acta vet. Scand. (suppl. 23): 48 pp.

LINDBERG, P. & JACOBSSON, S.O. (1970) Relationship between selenium content of forage, blood, and organs of sheep, and lamb mortality rate. Acta vet. Scand., 11: 49-58.

LIPINSKIJ, S. (1962) [Background for evaluation of selenium as an industrial toxicant.] Gig. i Sanit., 1: 91-93 (in Russian).

LO, L.W., KOROPATNICK, J., & STICH, H.F. (1978) The mutagenicity and cytotoxicity of selenite, "activated" selenite, and selenate for normal and DNA repair-deficient human fibroblasts. Mutat. Res., 49: 305-312.

LOFROTH, G. & AMES, B.N. (1978) Mutagenicity of inorganic compounds in Salmonella typhimurium: arsenic, chromium, and selenium. Mutat. Res., 53: 65-66.

LOMBECK, I., KASPEREK, K., FEINENDEGEN, L.E., & BREMER, H.J. (1975) Serum-selenium concentrations in patients with maple-syrup-urine disease and phenylketonuria under dieto-therapy. Clin. Chim. Acta, 64: 57-61.

LOMBECK, I., KASPEREK, K., HARBISCH, H.D., BECKER, K., SCHUMANN, E., SCHROTER, W., FEINENDEGEN, L.E., & BREMER, H.J. (1978) The selenium state of children. II. Selenium content of serum, whole blood, hair and the activity of erythrocyte glutathione peroxidase in dietetically treated patients with phenylketonuria and maple-syrup-urine disease. Eur. J. Pediatr., 128: 213-223.

LUO, X., WEI, H., YANG, C., XING, J., LIU, X., QIAO, C., FENG, Y., LIU, J., LIU, Y., WU, Q., LIU, X., GUO, J., STOECKER, B.J., SPALLHOLTZ, J.E., & YANG, S.P. (1985) Bioavailability of selenium to residents in a low-selenium area of China. Am. J. clin. Nutr., 42: 439-448.

MAAG, D.D. & GLENN, M.W. (1967) Toxicity of selenium: farm animals. In: Muth, O.H., Oldfield, J.E., & Weswig, P.H., ed. Selenium in biomedicine, Westport, Connecticut, The AVI Publishing Co., Inc, pp. 127-140.

MAAG, D.D., ORSBORN, J.S., & CLOPTON, J.R. (1960) The effect of sodium selenite on cattle. Am. J. vet. Res., 21: 1049-1053.

MCCABE, L.J., SYMONS, J.M., LEE, R.D., & ROBECK, G.G. (1970) Survey of community water supply systems. J. Am. Water Works Assoc., 62: 670-687.

MCCONNELL, K.P. & CHO, G.J. (1965) Transmucosal movement of selenium. Am. J. Physiol., 208: 1191-1195.

MCCONNELL, K.P. & PORTMAN, O.W. (1952a). Excretion of dimethyl selenide by the rat. J. Biol. Chem., 195: 277-282.

MCCONNELL, K.P. & PORTMAN, O.W. (1952b) Toxicity of dimethyl selenide in the rat and mouse. Proc. Soc. Exp. Biol. Med., 79: 230-231.

MCCONNELL, K.P., BROGHAMER, W.L., Jr, BLOTCKY, A.J., & HURT, O.J. (1975) Selenium levels in human blood and tissues in health and in disease. J. Nutr., 105: 1026-1031.

MCCONNELL, K.P., JAGER, R.M., BLAND, K.I., & BLOCTCKY, A.J. (1980) The relationship of dietary selenium and breast cancer. J. surg. Oncol., 15: 67-70.

MCCOY, K.E.M. & WESWIG, P.H. (1969) Some selenium responses in the rat not related to vitamin E. J. Nutr., 98: 383-389.

MCDANIEL, M., SHENDRIKAR, A.D., REISZNER, K.D., & WEST, P.W. (1976) Concentration and determination of selenium from environmental samples. Anal. Chem., 48: 2240-2243.

MCDOWELL, L.R., FROSETH, J.A., & PIPER, R.C. (1978) Influence of arsenic, sulfur, cadmium, tellurium, silver, and selenium on the selenium-vitamin E deficiency in the pig. Nutr. Rep. Int., 17: 19-33.

MCKEEHAN, W.L., HAMILTON, W.G., & HAM, R.G. (1976) Selenium is an essential trace nutrient for growth of WI-38 diploid human fibroblasts. Proc. Natl Acad. Sci. (USA), 73: 2023-2027.

MCKENZIE, J.M. (1977) Trace elements in total parenteral nutrition in New Zealand. In: Trace elements in human and animal health in New Zealand, Hamilton, New Zealand, Waikato University Press, pp. 59-69.

MCKENZIE, R.L., REA, H.M., THOMSON, C.D., & ROBINSON, M.F. (1978) Selenium concentration and glutathione peroxidase activity in blood of New Zealand infants and children. Am. J. clin. Nutr., 31: 1413-1418.

MACKINNON, A.M. & SIMON, F.R. (1976) Impaired hepatic heme synthesis in the phenobarbital-stimulated selenium-deficient rat. Proc. Soc. Exp. Biol. Med., 152: 568-572.

MCLEAN, A.E.M. (1967) The effect of diet and vitamin E on liver injury due to carbon tetrachloride. Br. J. exp. Path., 48: 632-636.

MAINES, M.D. & KAPPAS, A. (1976) Selenium regulation of hepatic heme metabolism: induction of δ-aminolevulinate synthase and heme oxygenase. Proc. Natl Acad. Sci.(USA), 73: 4428-4431.

MAJSTRUK, P.N. & SUCHKOV, B.P. (1978) [Sanitary hygienic control of selenium levels in the main food commodities and the daily intake of selenium,] Kiev, Kievskij, NII gig. Pit., 4 pp (in Russian).

MARSHALL, M.V., JACOBS, M.M., & GRIFFIN, A.C. (1978) Reduction in acetylaminofluorene (AAF) hepatocarcinogenesis by selenium. Proc. Am. Assoc. Cancer Res., 19: 75.

MARTIN, J.L. & GERLACH, M.L. (1969) Separate elution by ion-exchange chromatography of some biologically important selenoamino acids. Anal. Biochem., 29: 257-264.

MARTIN, J.L. & HURLBUT, J.A. (1976) Tissue selenium levels and growth responses of mice fed selenomethionine, Se-methylselenocysteine, or sodium selenite. Phosphorus Sulfur, 1: 295-300.

MARTIN, J.L. & SPALLHOLZ, J.S. (1976) Selenium in the immune response. In: Proceedings of the Symposium on Selenium-Tellurium in the Environment, Pittsburg, Pennsylvania, Industrial Health Foundation, pp. 204-225.

MARTIN, S.E., ADAMS, G.H., SCHILLACI, M., & MILNER, J.A. (1981) Antimutagenic effect of selenium on acridine orange and 7,12-dimethylbenz(α)anthracene in the Ames Salmonella microsomal system. Mutat. Res., 82: 41-46.

MASIRONI, R. & PARR, R. (1976) Selenium and cardiovascular diseases: preliminary results of the WHO/IAEA joint research programme. In: Proceeddings of the Symposium on Selenium-Tellurium in the Environment, Pittsburgh, Pennsylvania, Industrial Health Foundation, pp. 316-325.

MAXIA, V., MELONI, S., ROLLIER, M.A., BRANDONE, A., PATWARDHAN, V.N., WASLIEN, C.I., & SHAMI, S.E. (1972) Selenium and chromium assay in Egyptian foods and in blood of Egyptian children by activation analysis. In: Nuclear activation techniques in the life sciences, Vienna, International Atomic Energy Agency, pp. 527-550.

MEDINA, D. & SHEPHERD, F. (1980) Selenium-mediated inhibition of mouse mammary tumorigenesis. Cancer Lett., 8: 241-245.

MEDINSKY, M.A., CUDDIHY, R.G., GRIFFITH, W.C., & MCCLELLAN, R.O. (1981) A simulation model describing the metabolism of inhaled and ingested selenium compounds. Toxicol. appl. Pharmacol., 59: 54-63.

MEHLERT, A. & DIPLOCK, A.T. (1985) The glutathione-S-transferases in selenium and vitamin E deficiency. Biochem. J., 227: 823-831.

MICHIE, N.D., DIXON, E.J., & BUNTON, N.G. (1978) Critical review of AOAC fluorometric method for determining selenium in foods. J. Assoc. Off. Anal. Chem., 61: 48-51.

MIETTINEN, T.A., ALFTHAN, G., HUTTUNEN, J.K., PIKKARAINEN, J., NAUKKARINEN, V., MATTILA, S., & KUMLIN, T. (1983) Serum selenium concentration related to myocardial infarction and fatty acid content of serum lipids. Brit. med. J., 287: 517-519.

MILKS, M.M., WILT, S.R., ALI, I.I., & COURI, D. (1985) The effects of selenium on the emergence of aflatoxin B_1-induced enzyme-altered foci in rat liver. Fundam. appl. Toxicol., 5: 320-326.

MILLAR, K.R. & SHEPPARD, A.D. (1972) α-Tocopherol and selenium levels in human and cow's milk. N.Z. J. Sci., 15: 3-15.

MILLAR, K.R., GARDINER, M.A., & SHEPPARD, A.D. (1973) A comparison of the metabolism of intravenously injected sodium selenite, sodium selenate, and selenomethionine in rats. N.Z. J. agric. Res., 16: 115-127.

MILLER, D., SOARES, J.H., Jr, BAUERSFELD, P., Jr, & CUPPETT, S.L. (1972) Comparative selenium retention by chicks fed sodium selenite, selenomethionine, fish meal, and fish solubles. Poultry Sci., 51: 1669-1673.

MILNER, J.A. (1985) Effect of selenium on virally induced and transplantable tumour models. Fed. Proc., 44: 2568-2572.

MONAENKOVA, A.M. & GLOTOVA, K.V. (1963) [Selenium intoxication.] Gig. i Sanit., 6: 41-44 (in Russian).

MONDRAGON, M.C. & JAFFE, W.G. (1976) Ingestion of selenium in Caracas, compared with some other cities. Arch. Latinoam. Nutr., 26: 341-352.

MONEY, D.F.L. (1970) Vitamin E and selenium deficiencies and their possible aetiological role in the sudden death in infants syndrome. N.Z. med. J., 71: 32-34.

MONEY, D.F.L. (1978) Vitamin E, selenium, iron and vitamin A content of livers from sudden infant death syndrome and control children: interrelationships and possible significance. N.Z. J. Sci Teel., 21: 41-55.

MOORE, J.A., NOIVA, R., & WELLS, I.C. (1984) Selenium concentrations in plasma of patients with arteriographically defined coronary atherosclerosis. Clin. Chem., 30(7): 1171-1173.

MORIYA, M., OHTA, T., WATANABE, K., WATANABE, Y., SUGIYAMA, F., MIYAZAWA, T., & SHIRASU, Y. (1979) Inhibitors for the mutagenicities of colon carcinogens, 1,2-dimethylhydrazine and azoxymethane, in the host-mediated assay. Cancer Lett., 7: 325-330.

MORRIS, V.C. & LEVANDER, O.A. (1970) Selenium content of foods. J. Nutr., 100: 1383-1388.

MOTSENBOCKER, M.A. & TAPPEL, A.L. (1982) A selenocysteine-containing selenium-transport protein in rat plasma. Biochim. Biophys. Acta, 719: 147-153.

MOXON, A.L. (1938) The effect of arsenic on the toxicity of seleniferous grains. Science, 88: 81.

MOXON, A.L. (1940) Toxicity of selenium-cystine and some other organic selenium compounds. J. Am. Pharm. Assoc. Sci. Ed., 29: 249-250.

MOXON, A.L. (1976) Natural occurrence of selenium. In: Proceedings of the Symposium on Selenium-Tellurium in the Environment, Pittsburgh, Pennsylvania, Industrial Health Foundation, pp. 1-9.

MOXON, A.L. & RHIAN, M. (1943) Selenium poisoning. Physiol. Rev., 23: 305-337.

MOXON, A.L., ANDERSON, H.D., & PAINTER, E.P. (1938) The toxicity of some organic selenium compounds. J. Pharmacol. exp. Ther., 63: 357-368.

MOXON, A.L., OLSON, O.E., & SEARIGHT, W.V. (1939) Selenium in rocks, soils, and plants. S. Dak. Agric. Exp. Sta. Tech. Bull., 2: 94 pp.

MOXON, A.L., SCHAEFER, A.E., LARDY, H.A., DUBOIS, K.P., & OLSON, O.E. (1940) Increasing the rate of excretion of selenium from selenized animals by the administration of p-bromobenzene. J. biol. Chem., 132: 785-786.

MOXON, A.L., OLSON, O.E., & SEARIGHT, W.V. (1950) Selenium in rocks, soils, and plants. S. Dak. Agric. Exp. Sta. Tech. Bull., 2.

MUHLER, J.C. & SHAFER, W.G. (1957) The effect of selenium on the incidence of dental caries in rats. J. dent. Res., 36: 895-896.

MUNSELL, H.E., DEVANEY, G.M., & KENNEDY, M.H. (1936) Toxicity of food containing selenium as shown by its effect on the rat, Washington DC, US Department of Agriculture, 25 pp (US Department of Agriculture Technical Bulletin No. 534).

MUTANEN, M. & KOIVISTOINEN, P. (1983) The role of imported grain on the selenium intake of Finnish population 1941-1981. Int. J. Vit. Nutr. Res., 53: 34-38.

MUTH, O.H. (1960) Carbon tetrachloride poisoning of ewes on a low selenium ration. Am. J. vet. Res., 21: 86-87.

MUTH, O.H., ed. (1966) Selenium in biomedicine. In: Proceedings of the First International Symposium at Oregon State University, Westport, Connecticut, AVI Publishing Co., 445 pp.

MUTH, O.H., WESWIG, P.H., WHANGER, P.D., & OLDFIELD, J.E. (1971) Effect of feeding selenium-deficient ration to the subhuman primate (Saimiri sciureus). Am. J. vet. Res., 32: 1603-1605.

NAHAPETIAN, A.T., JANGHORBANI, M., & YOUNG, V.R. (1983) Urinary trimethylselenonium excretion by the rat: effect of level and source of selenium - 75. J. Nutr., 113: 401-411.

NAKAMURO, K., YOSHIKAWA, K., SAYATO, Y., KURATA, H., TONOMURA, M., & TONOMURA, A. (1976) Studies on selenium-related compounds. V. Cytogenetic effect and reactivity with DNA. Mutat. Res., 40: 177-184.

NATIONAL CANCER INSTITUTE (1980) Bioassay of selenium sulfide for possible carcinogenicity (gavage study), Bethesda, Maryland, US Department of Health and Human Services, Public Health Service, National Institutes of Health, 132 pp (DHHS Publication No. (NIH) 80-1750).

NAVIA, J.M., MENAKER, L., SELTZER, J., & HARRIS, R.S. (1968) Effect of Na$_2$SeO$_3$ supplemented in the diet or the water on dental caries of rats. Fed. Proc., 27: 676.

NAZARENKO, I.I. & ERMAKOV, A.N. (1971) Analytical chemistry of selenium and tellurium, New York, Halsted Press.

NAZARENKO, I.I. & KISLOVA, I.V. (1977) [A highly sensitive method for the analysis of migratory forms of selenium in natural waters.] El. Viems. Labor. Technol. issled. obogashch. min. syrja, 6: 1-8 (in Russian).

NAZARENKO, I.I., KISLOVA, I.V., GUSEJNOV, T.M., MKRTCHJAN, M.A., & KISLOV, A.M. (1975) [Fluorometric determination of selenium in biological material by 2,3-diaminonaphthalene]. Zh. Anal. Chim., 30(4): 733-737 (in Russian).

NELSON, A.A., FITZHUGH, O.G., & CALVERY, H.O. (1943) Liver tumors following cirrhosis caused by selenium in rats. Cancer Res., 3: 230-236.

NEWBERNE, P.M. & CONNER, M.W. (1974) Effect of selenium on acute response to aflatoxin B$_1$. In: Hemphill, D.D., ed. Trace substances in environmental health. VIII, Columbia, Missouri, University of Missouri Press, pp. 323-328.

NOCKELS, C.F. (1979) Protective effects of supplemental vitamin E against infection. Fed. Proc., 38: 2134-2138.

NODA, M., TAKANO, T., & SAKURAI, H. (1979) Mutagenic activity of selenium compounds. Mutat. Res., 66: 175-179.

NOGUCHI, T., CANTOR, A.H., & SCOTT, M.L. (1973) Mode of action of selenium and vitamin E in prevention of exudative drathesis in chicks. J. Nutr., 103(10): 1502-1511.

NORDMAN, E. (1974) Sodium selenite (selenium-75) scintigraphy in diagnosis of tumours. Acta radiol., 340(suppl.): 78 pp.

NORPPA, H., ET AL. (1980) Chromosomal effects of sodium selenite in vivo. Hereditas, 93: 93-105.

NORRIS, F.H. & SANG, K. (1978) Amyotrophic lateral sclerosis and low urinary selenium levels. J. Am. Med. Assoc., 239: 404.

OBERMEYER, B.D., PALMER, I.S., OLSON, O.E., & HALVERSON, A.W. (1971) Toxicity of trimethylselenonium chloride in the rat with and without arsenite. Toxicol. appl. Pharmacol., 20: 135-146.

OCHOA-SOLANO, A. & GITLER, C. (1968) Incorporation of ^{75}Se-selenomethionine and ^{35}S-methionine into chicken egg white proteins. J. Nutr., 94: 243-248.

OELSCHLAGER, W. & MENKE, K.H. (1969) Concerning the selenium content of plant, animal, and other materials. II. The selenium and sulfur content of foods. Z. Ernahrungswiss., 9: 216-222.

OH, S.H., SUNDE, R.A., POPE, A.L., & HOEKSTRA, W.G. (1976a) Glutathione peroxidase response to selenium intake in lambs fed a Torula yeast-based, artificial milk. J. anal. Sci., 42: 977-983.

OH, S.H., POPE, A.L., & HOEKSTRA, W.G. (1976b) Dietary selenium requirement of sheep fed a practical-type diet as assessed by tissue glutathione peroxidase and other criteria. J. anal. Sci., 42: 984-992.

OHLENDORF, H.M., HOFFMAN, D.J., SAIKI, M.K., & ALDRICH, T.W. (1986) Embryonic mortality and abnormalities of aquatic birds: apparent impact of selenium from irrigation drain water. Sci. total Environ., 52: 49-63.

OLSON, O.E. (1967) Soil, plant, animal cycling of excessive levels of selenium. In: Muth, O.H., Oldfield, J.E., & Weswig, P.H., ed. Selenium in biomedicine, Westport, Connecticut, The AVI Publishing Co., Inc, pp. 297-312.

OLSON, O.E. (1969) Selenium as a toxic factor in animal nutrition. In: Proceedings of the Georgia Nutrition Conference, Atlanta, Georgia, pp. 68-78.

OLSON, O.E. (1970) Selenium in feedstuffs: deficiencies and excesses. Proceedings of the 31st Minnesota Nutrition Conference, Minneapolis, University of Minnesota, pp. 7-13.

OLSON, O.E. (1976) Methods of analysis for selenium. A review. In: Proceedings of the Symposium on Selenium-Tellurium in the Environment, Pittsburgh, Pennsylvania, Industrial Health Foundation, pp. 67-84.

OLSON, O.E. (1978) Selenium in plants as a cause of livestock poisoning. In: Keeler, R.F., Van Kampen, K.R., &

James, L.F., ed. Effects of poisonous plants on livestock, New York, Academic Press, pp. 121-133.

OLSON, O.E. & FROST, D.V. (1970) Selenium in papers and tabaccos. Environ. Sci. Technol., 4: 686-687.

OLSON, O.E. & PALMER, I.S. (1976) Selenoamino acids in tissues of rats administered inorganic selenium. Metabolism, 25: 299-306.

OLSON, O.E., WHITEHEAD, E.I., & MOXON, A.L. (1942) Occurrence of soluble selenium in soils and its availability to plants. Soil Sci., 54: 47-53.

OLSON, O.E., SCHULTE, B.M., WHITEHEAD, E.I., & HALVERSON, A.W. (1963) Effect of arsenic on selenium metabolism in rats. J. agric. food Chem., 11: 531-534.

OLSON, O.E., NOVACEK, E.J., WHITEHEAD, E.I., & PALMER, I.S. (1970) Investigations on selenium in wheat. Phytochemistry, 9: 1181-1188.

OLSON, O.E., PALMER, I.S., & WHITEHEAD, E.I. (1973) Determination of selenium in biological materials. In: Glick, D., ed. Methods of biochemical analysis, New York, John Wiley and Sons, Inc, Vol. 21, pp. 39-78.

OLSON, O.E., PALMER, I.S., & CARY, E.E. (1975) Modification of the official fluorometric method for selenium in plants. J. Assoc. Off. Anal. Chem., 58: 117-121.

OLSON, O.E., CARY, E.E., & ALLAWAY, W.H. (1976) Fixation and volatilization by soils of selenium from trimethylselenonium. Agron. J., 68: 839-843.

OLSON, O.E., PALMER, I.S., & HOWE, M. (1978) Selenium in foods consumed by South Dakotans. Proc. S.D. Acad. Sci., 57: 113-121.

OLSSON, U., ONFELT, A., & BEIJE, B. (1984) Dietary selenium deficiency causes decreased N-oxygenation of N,N-dimethyl-aniline and increased mutagenicity of dimethylnitrosamine in the isolated rat liver/cell cultrue system. Mutat. Res., 126: 73-80.

OMAYE, S.T., REDDY, K.A., & CROSS, C.E. (1978) Enhanced lung toxicity of paraquat in selenium-deficient rats. Toxicol. appl. Pharmacol., 43: 237-247.

OSTADALOVA, I., BABICKY, A., & OBENBERGER, J. (1979) Cataractogenic and lethal effect of selenite in rats during postnatal ontogenesis. Physiol. Bohemoslov., 28: 393-397.

OVERVAD, K., THORLING, E.B., BJERRING, P., & EBBESEN, P. (1985) Selenium inhibits UV-light-induced skin carcinogenesis in hairless mice. Cancer Lett., 27: 163-170.

PAINTER, E.P. (1941) The chemistry and toxicity of selenium compounds with special reference to the selenium problem. Chem. Rev., 28: 179-213.

PALMER, I.S. & OLSON, O.E. (1974) Relative toxicities of selenite and selenate in the drinking-water of rats. J. Nutr., 104(3): 306-314.

PALMER, I.S. & OLSON, O.E. (1979) Partial prevention by cyanide of selenium poisoning in rats. Biochem. Biophys. Res. Commun., 90: 1379-1386.

PALMER, I.S., FISCHER, D.D., HALVERSON, A.W., & OLSON, O.E. (1969) Identification of a major selenium excretory product in rat urine. Biochim. Biophys. Acta, 177: 336-342.

PALMER, I.S., GUNSALUS, R.P., HALVERSON, A.W., & OLSON, O.E. (1970) Trimethylselenonium ion as a general excretory product from selenium metabolism in the rat. Biochim. Biophys. Acta, 208: 260-266.

PALMER, I.S., ARNOLD, R.L., & CARLSON, C.W. (1973) Toxicity of various selenium derivatives to chick embryos. Poultry Sci., 52: 1841-1846.

PALMER, I.S., OLSON, O.E., HALVERSON, A.W., MILLER, R., & SMITH, C. (1980) Isolation of factors in linseed oil meal protective against chronic selenosis in rats. J. Nutr., 110: 145-150.

PARIZEK, J. & OSTADALOVA, I. (1967) The protective effect of small amounts of selenite in sublimate intoxication. Experientia (Basel), 23: 142-143.

PARIZEK, J., OSTADALOVA, I., KALOUSKOVA, J., BABICKY, A., & BENES, J. (1971) The detoxifying effects of selenium. Interrelations between compounds of selenium and certain metals. In: Mertz, W. & Cornatzer, W.E., ed. Newer trace elements in nutrition, New York, Marcel Dekker, Inc, pp. 85-122.

PARIZEK, J., ET AL. (1974) Some hormonal and environmental factors influencing selenium metabolism and action. In: Proceedings of the Ninth International Congress in Nutrition, Mexico, 1972 (Coop. Nutr. Cong., 16: 94-97).

PARIZEK, J., KALOUSKOVA, J., KORUNOVA, V., BENES, J., & PAVLIK, L. (1976) The protective effect of pretreatment with selenite on the toxicity of dimethylselenide. Physiol. Bohemoslov., 25: 573-576.

PARIZEK, J., KALOUSKOVA, J., BENES, J., & PAVLIK, L. (1980) Interactions of selenium-mercury and selenium-selenium compounds. Ann. N.Y. Acad. Sci., 355: 347-360.

PASCOE, G.A. & CORREIA, M.A. (1985) Structural and functional assembly of rat intestinal cytochrome P-450 isozymes: effects of dietary iron and selenium. Biochem. Pharmacol., 34: 599-608.

PASCOE, G.A., SAKAI-WONG, J., SOLIVEN, E., & CORREIA, M.A. (1983) Regulation of intestinal cytochrome P-450 and Heme by dietary nutrients. Critical role of selenium. Biochem. Pharmacol., 32(20): 3027-3035.

PAULSEN, P.J. (1977) Determination of cadmium, copper, iron, lead, mercury, molybdenum, nickel, selenium, silver, tellurium, thallium, and zinc. Natl Bur. Stand. Special Publ. (USA), 492: 33-48.

PAVLIK, L., KALOUSKOVA, J., VOBECKY, M., DEDINA, BENES, J., & PARIZEK, J. (1979) Selenium levels in the kidneys of male and female rats. In: Nuclear activation techniques in the life sciences, 1978, Vienna, International Atomic Energy Agency, pp. 213-223.

PELEKIS, E.E., PELEKIS, L.L., & TAURE, I. JA. (1975) [Instrumental neutron-activation method of selenium determination in biological materials.] Vitaminy, 8: 146-150 (in Russian).

PENCE, B.C. & BUDDINGH, F. (1985) Effect of selenium deficiency on incidence and size of 1,2-dimethylhydrazine-induced colon cancer in rats. J. Nutr., 115: 1196-1202.

PENNINGTON, J.A.T., WILSON, D.B., NEWELL, R.F., HARLAND, B.F., JOHNSON, R.D., & VANDERVEEN, J.E. (1984) Selected minerals in foods surveys, 1974 to 1981/82. J. Am. Diet. Assoc., 84: 771-780.

PERONA, G., GUIDI, G.C., PIGA, A., CELLERINO, R., MENNA, R., & ZATTI, M. (1978) In vivo and in vitro variations of human erythrocyte glutathione peroxidase activity as result of cells ageing, selenium availability and peroxide activation. Br. J. Haematol., 39: 399-408.

PERRY, H.M., Jr, ERLANGER, M.W., & PERRY, E.F. (1976) Limiting conditions for the induction of hypertension in rats by cadmium. In: Hemphill, D.D., ed. Trace substances in environmental health. X, Columbia, Missouri, University of Missouri Press, pp. 459-467.

PIERCE, F.D. & BROWN, H.R. (1977) Comparison of inorganic interferences in atomic absorption spectrometric determination of arsenic and selenium. Anal. Chem., 49: 1417-1422.

PIERCE, S. & TAPPEL, A.L. (1977) Effects of selenite and selenomethionine on glutathione peroxidase in the rat. J. Nutr., 107: 475-479.

PILLAY, K.K.S., THOMAS, C.C., Jr, & SONDEL, J.A. (1971) Activation analysis of airborne selenium as a possible indicator of atmospheric sulfur pollutants. Environ. Sci. Technol., 5: 74-77.

PLAA, G.L. & WITSCHI, H. (1976) Chemicals, drugs, and lipid peroxidation. Ann. Rev. Pharmacol. Toxicol., 16: 125-141.

PLEBAN, P.A., MUNYANI, A., & BEACHUM, J. (1982) Determination of selenium concentration and glutathine peroxidase activity in plasma and erythrocytes. Clin. Chem., 28: 311-316.

PLETNIKOVA, I.P. (1970) Biological effect and safe concentration of selenium in drinking-water. Hyg. Sanit., 35: 176-181.

POLEY, W.E. & MOXON, A.L. (1938) Tolerance levels of seleniferous grains in laying rations. Poultry Sci., 17: 72-76.

POOLE, C.F., EVANS, N.J., & WIBBERLEY, D.G. (1977) Determination of selenium in biological samples by gas-liquid chromatography with electron-capture detection. J. Chromatogr., 136: 73-83.

PRATLEY, J.E. & MCFARLANE, J.D. (1974) The effect of sulphate on the selenium content of pasture plants. Aust. J. exp. Agric. Anim. Husb., 14: 533-538.

PRINGLE, P. (1942) Occupational dermatitis following exposure to inorganic selenium compounds. Br. J. Dermatol. Syphil., 54: 54-58.

PROHASKA, J.R. & GANTHER, H.E. (1976) Selenium and glutathione peroxidase in developing rat brain. J. Neurochem., 27: 1379-1387.

PROHASKA, J.R. & GANTHER, H.E. (1977) Glutathione peroxidase activity of glutathione-S-transferases purified from rat liver. Biochem. Biophys. Res. Commun., 76: 437-445.

PYEN, G. & FISHMAN, M. (1978) Automated determination of selenium in water. At. Absorpt. Newslett., 17: 47-48.

RANDOLPH, W.F. (1978) Food additives permitted in feed and drinking water of animals. Selenium. Fed. Reg., 43: 11700-11701.

RANSONE, J.W., SCOTT, N.M., Jr, & KNOBLOCK, E.C. (1961) Selenium sulfide intoxication. New Engl. J. Med., 264: 384-385.

RAVIKOVITCH, S. & MARGOLIN, M. (1957) Selenium in soils and plants. Ktavim. Rec. agric. Res. Sta., 7: 41-52.

RAY, J.H. (1984) Sister-chromatid exchange induction by sodium selenite: reduced glutathione converts Na_2SeO_3 to its SCE-inducing form. Mutat. Res., 141: 49-53.

RAY, J.H. & ALTENBURG, L.C. (1978) Sister-chromatid exchange induction by sodium selenite: dependence on the presence of red blood cells or red blood cell lysate. Mutat. Res., 54: 343-354.

RAY, J.H. & ALTENBURG, L.C. (1980) Dependence of the sister-chromatid exchange-inducing abilities of inorganic selenium compounds on the valence state of selenium. Mutat. Res., 78: 261-266.

RAY, J.H., ALTENBURG, L.C., & JACOBS, M.M. (1978) Effect of sodium selenite and methyl methanesulfonate or N-hydroxy-2-acetylamino-fluorene co-exposure on sister-chromatid exchange production in human whole blood cultures. Mutat. Res., 57: 359-368.

REA, H.M., THOMSON, C.D., CAMPBELL, D.R., & ROBINSON, M.F. (1979) Relation between erythrocyte selenium concentrations and glutathione peroxidase (EC 1.11.1.9) activities of New

Zealand residents and visitors to New Zealand. Br. J. Nutr.,
42: 201-208.

REAMER, D.C. & VEILLON, C. (1983a) Letter to the editor.
Anal. Chem., in press.

REAMER, D.C. & VEILLON, C. (1983b) A double isotope dilution
method for using stable selenium isotapes in metabolic tracer
studies: analysis by gas chromatography/mass spectrometry
(GC/MS). J. Nutr., 113: 786-792.

REITER, R. & WENDEL, A. (1983) Selenium and drug metabolism.
Multiple modulations of mouse liver enzymes. Biochem.
Pharmacol., 32(20): 3063-3067.

RHEAD, W.J., CARY, E.E., ALLAWAY, W.H., SALTZSTEIN, S.L., &
SCHRAUZER, G.N. (1972) The vitamin E and selenium status of
infants and the sudden infant death syndrome. Bioinorg. Chem.,
1: 289-294.

RICHOLD, M., ROBINSON, M.F., & STEWART, R.D.H. (1977)
Metabolic studies in rats of ^{75}Se incorporated in vivo into
fish muscle. Br. J. Nutr., 38: 19-29.

RIELY, C.A., COHEN, G., & LIEBERMAN, M. (1974) Ethane
evolution: a new index of lipid peroxidation. Science, 183:
208-210.

RILEY, J.F. (1968) Mast cells, co-carcinogenesis and
anti-carcinogenesis in the skin of mice. Experientia (Basel),
24: 1237-1238.

ROBBINS, C.W. & CARTER, D.L. (1970) Selenium concentrations
in phosphorus fertilizer materials and associated uptake by
plants. Soil Sci. Soc. Am. Proc., 34: 506-509.

ROBERTSON, D.S.F. (1970) Selenium, a possible teratogen?
Lancet, 1: 518-519.

ROBINSON, J.R., ROBINSON, M.F., LEVANDER, O.A., & THOMSON,
C.D. (1985) Urinary excretion of selenium by New Zealand and
North American human subjects on differing intakes. Am. J.
clin. Nutr., 41: 1023-1031.

ROBINSON, M.F. & THOMSON, C.D. (1981) Selenium levels in
humans vs environmental sources. In: Spallholz, J.E., Martin,
J.L., & Ganther, H.E., ed. Selenium in biology and medicine,
Westport, Connecticut, The AVI Publishing Co., Inc, pp. 283-302

ROBINSON, M.F. & THOMSON, C.D. (1983) The role of selenium in the diet. Nutr. Abstr. Rev., 53: 3-26.

ROBINSON, M.F. & THOMSON, C.D. (1986) Selenium status of the food supply and residents of New Zealand (Beijing Conferenc).

ROBINSON, M.F., MCKENZIE, J.M., THOMSON, C.D., & VAN RIJ, A.L. (1973) Metabolic balance of zinc, copper, cadmium, iron, molybdenum and selenium in young New Zealand women. Br. J. Nutr., 30: 195-205.

ROBINSON, M.F., GODFREY, P.J., THOMSON, C.D., REA, H.M., & VAN RIJ, A.M. (1978a) Blood selenium and glutathione peroxidase activity in normal subjects and in surgical patients with and without cancer in New Zealand. Am. J. clin. Nutr., 32: 1477-1485.

ROBINSON, M.F., REA, H.M., FRIEND, G.M., STEWART, R.D.H., SNOW, P.C., & THOMSON, C.D. (1978b) On supplementing the selenium intake of New Zealanders. II. Prolonged metabolic experiments with daily supplements of selenomethionine, selenite, and fish. Br. J. Nutr., 39: 589-600.

ROBINSON, M.F., GODFREY, J.P., THOMSON, C.D., REA, H.M., & VAN RIJ, A.M. (1979) Blood selenium and glutathione peroxidase activity in normal subjects and in surgical patients with and without cancer in New Zealand. Am. J. clin. Nutr., 32: 1477-1485.

ROBINSON, M.F., CAMPBELL, D.R., STEWART, R.D.H., REA, H.M., THOMSON, C.D., SNOW, P.G., & SQUIRES, I.H.W. (1981) Effect of daily supplements of selenium on patients with muscular complaints in Otago and Canterbury. N.Z. med. J., 93: 289-292.

ROBINSON, M.F., CAMPBELL, D.R., SUTHERLAND, W.H.F., HERBISON, P.G., PAULIN, J.M., & SIMPSON, F.O. (1983) Selenium and risk factors for cardiovascular disease in New Zealand. N.Z. med. J., 96: 755-757.

ROBINSON, M.F., THOMSON, C.D., & HUEMMER, P.K. (1985) Effect of a megadose of ascorbic acid, a meal and orange juice on the absorption of selenium as sodium selenite. N.Z. med. J., 98: 627-629.

ROBINSON, W.O. (1933) Determination of selenium in wheat and soils. J. Assoc. Off. Agric. Chem., 16: 423-424.

ROHMER, R., CARROT, E., & GOUFFAULT, J. (1950) Nouvel aspect de l'intoxication par les composés du sélénium. Bull. Soc. Chim. France, 5(17): 275-278.

ROSIN, M.P. & STICH, H.F. (1979) Assessment of the use of the Salmonella mutagenesis assay to determine the influence of antioxidants on carcinogen-induced mutagenesis. Int. J. Cancer, 23: 722-727.

ROSENFELD, I. & BEATH, O.A. (1954) Effect of selenium on reproduction in rats. Proc. Soc. Exp. Biol. Med., 87: 295-297.

ROSENFELD, I. & BEATH, O.A. (1964) Selenium geobotany, biochemistry, toxicity and nutrition, New York, Academic Press, 411 pp.

ROSIN, M.P. (1981) Inhibition of spontaneous mutagenesis in yeast cultures by selenite, selenate, and selenide. Cancer Lett., 13: 7-14.

ROSSI, L.C., CLEMENTE, G.F., & SANTARONI, G. (1976) Mercury and selenium distribution in a defined area and in its population. Arch. environ. Health, 31: 160-165.

ROTRUCK, J.T., POPE, A.L., GANTHER, H.E., SWANSON, A.B., HAFEMAN, D.G., & HOEKSTRA, W.G. . (1973) Selenium: biochemical role as a component of glutathione peroxidase. Science, 179: 588-590.

RUDERT, C.P. & LEWIS, A.R. (1978) The effect of potassium cyanide on the occurrence of nutritional myopathy in lambs. Rhod. J. agric. Res., 16: 109-116.

RUSIECKI, W. & BRZEZINSKI, J. (1966) Influence of sodium selenate on acute thallium poisonings. Acta Pol. pharm., 23: 74-80.

SAKURAI, H. & TSUCHIYA, K. (1975) A tentative recommendation for the maximum daily intake of selenium. Environ. Physiol. Biochem., 5: 107-118.

SALONEN, J.T. (1985) Selenium in cardiovascular diseases and cancer. Epidemiologic findings from Finland. In: Boström, H. & Ljungstedt, N., ed. Trace elements in health disease, Stockholm, Sweden, Almqvist & Wiksell International, pp. 172-186.

SALONEN, J.T., ALFTHAN, G., HUTTUNEN, J.K., PIKKARAINEN, J., & PUSKA, P. (1982) Association between cardiovascular death and myocardial infarction and serum selenium in a watched-pair longitudinal study. Lancet, 2: 175-179.

SALONEN, J.T., ALFTHAN, G., HUTTUNEN, J.K., & PUSKA, P. (1984a) Association between serum selenium and the risk of cancer. Am. J. Epidemiol., 120(3): 342-349.

SALONEN, J.T., SALONEN, R., PENTTILA, I., HERRANEN, J., JAUHI-AINEN, J., KANTOLA, M., KAURANEN, P., LAPPETELAINEN, R., MAEN-PAA, P., & PUSKA, P. (1984b) Serum fatty acids, apolipoproteins, selenium and vitamin antioxidants and the risk of death from ischaemic heart disease. Am. J. Cardiol., 56(2): 226.

SALONEN, J.T., SALONEN, R., LAPPETELAINEN, R., MAENPAA, P.H., ALFTHAN, G., & PUSKA, P. (1985) Risk of cancer in relation to serum concentrations of selenium and vitamins A and E: matched case-control analysis of prospective data. Brit. med. J., 290: 417-420.

SARATHCHANDRA, S.U. & WATKINSON, J.H. (1981) Oxidation of elemental selenium to selenite by Bacillus megaterium. Science, 211: 600-601.

SARATHCHANDRASCHMIDT, K. & HELLER, W. (1976) Selenium concentration and activity of glutathione peroxidase in lysates of human erythrocytes. Blut, 33: 247-251.

SCHECTER, A., SHANSKE, W., STENZLER, A., QUINTILIAN, H., & STEINBERG, H. (1980) Acute hydrogen selenide intoxication. Chest, 77: 554-555.

SCHILLACI, M., MARTIN, S.E., & MILNER, J.A. (1982) The effects of dietary selenium on the biotransformation of 7,12-dimethylbenz(α)anthracene. Mutat. Res., 101: 31-37.

SCHRAUZER, G.N. (1976) Cancer mortality correlation studies. II. Regional associations of mortalities with the consumptions of foods and other commodities. Medic. Hypoth., 2: 39-49.

SCHRAUZER, G.N. & ISHMAEL, D. (1974) Effects of selenium and of arsenic on the genesis of spontaneous mammary tumors in inbred C3H mice. Ann. clin. Lab. Sci., 4: 441-447.

SCHRAUZER, G.N. & WHITE, D.A. (1978) Selenium in human nutrition: dietary intakes and effects of supplementation. Bioinorg. Chem., 8: 303-318.

SCHRAUZER, G.N., WHITE, D.A., & SCHNEIDER, C.J. (1976) Inhibition of the genesis of spontaneous mammary tumors in

C₃H mice: effects of selenium and of selenium-antagonistic elements and their possible role in human breast cancer. Bioinorg. Chem., 6: 265-270.

SCHRAUZER, G.N., WHITE, D.A., & SCHNEIDER, C.J. (1977) Cancer mortality correlation studies. III. Statistical associations with dietary selenium intakes. Bioinorg. Chem., 7: 23-34.

SCHRAUZER, G.N, WHITF, D.A., & SCHNEIDER, C.J. (1978a) Effects of selenium, arsenic, and zinc on the genesis of spontaneous mammary tumors in inbred female C₃H mice. In: Kirchgessner, M., ed. Trace element metabolism in man and animals. III, Freising-Weihenstephan, West Germany, Arbeitskreis für Tierernährungsforschung Weihenstephan, pp. 387-390.

SCHRAUZER, G.N., WHITE, D.A., & SCHNEIDER, C.J. (1978b) Selenium and cancer: Effects of selenium and of the diet on the genesis of spontaneous mammary tumors in virgin inbred female C₃H/St mice. Bioinorg. Chem., 8: 387-396.

SCHRAUZER, G.N., WHITE, D.A., MCGINNESS, J.E., SCHNEIDER, C.J., & BELL, L.J. (1978c) Arsenic and cancer: effects of joint administration of arsenite and selenite on the genesis of mammary adenocarcinoma in inbred female C₃H/St mice. Bioinorg. Chem., 9: 245-253.

SCHROEDER, H.A. & MITCHENER, M. (1971a) Toxic effects of trace elements on the reproduction of mice and rats. Arch. environ. Health, 23: 102-106.

SCHROEDER, H.A. & MITCHENER, M. (1971b) Selenium and tellurium in rats: effect on growth, survival and tumors. J. Nutr., 101: 1531-1540.

SCHROEDER, H.A. & MITCHENER, M. (1972) Selenium and tellurium in mice. Effects on growth, survival, and tumors. Arch. environ. Health, 24: 66-71.

SCHROEDER, H.A., FROST, D.V., & BALASSA, J.J. (1970) Essential trace metals in man: selenium. J. chron. Dis., 23: 227-243.

SCHUBERT, J.R., MUTH, O.H., OLDFIELD, J.E., & REMMERT, L.F. (1961) Experimental results with selenium in white muscle disease of lambs and calves. Fed. Proc., 20: 689-694.

SCHULTZ, T.D. & LEKLEM, J.E. (1983) Selenium status of vegetarians, nonvegetarians, and hormone-dependent cancer subjects. Am. J. clin. Nutr., 37: 114-118.

SCHWARZ, K. (1961) Development and status of experimental work on Factor 3-selenium. Fed. Proc., 20: 666-673.

SCHWARZ, K. (1962) Vitamin E, trace elements, and sulfhydryl groups in respiratory decline. Vitam. Horm., 20: 463-484.

SCHWARZ, K. (1967) Discussion comments. In: Muth, O.H., Oldfield, J.E., & Weswig, P.H., ed. Selenium in biomedicine, Westport, Connecticut, The AVI Publishing Co., Inc, pp. 225-226.

SCHWARZ, K. (1976) The discovery of the essentiality of selenium, and related topics (a personal account). In: Proceedings of the Symposium on Selenium-Tellurium in the Environment, Pittsburgh, Pennsylvania, Industrial Health Foundation, pp. 349-376.

SCHWARZ, K. (1977) Amyotrophic lateral sclerosis and selenium. J. Am. Med. Assoc., 238: 2365.

SCHWARZ, K. & FOLTZ, C.M. (1957) Selenium as an integral part of Factor 3 against dietary necrotic liver degeneration. J. Am. Chem. Soc., 79: 3292-3293.

SCHWARZ, K. & FOLTZ, C.M. (1958) Factor 3 activity of selenium compounds. J. biol. Chem., 233: 245-251.

SCHWARZ, K., PORTER, L.A., & FREDGA, A. (1972) Some regularities in the structure-function relationship of organoselenium compounds effective against dietary liver necrosis. Ann. NY Acad. Sci., 192: 200-214.

SCOTT, M.L. & THOMPSON, J.N. (1971) Selenium content of feedstuffs and effects of dietary selenium levels upon tissue selenium in chicks and poults. Poultry Sci., 50: 1742-1748.

SCOTT, M.L., OLSON, G., KROOK, L., & BROWN, W.R. (1967) Selenium-responsive myopathies of myocardium and of smooth muscle in the young poult. J. Nutr., 91: 573-583.

SEIFTER, J., EHRICH, W.E., HUDYMA, G., & MUELLER, G. (1946) Thyroid adenomas in rats receiving selenium. Science, 103: 762.

SELJANKINA, K.P., JAHIMOVICH, N.P., ALEKSEEVA, L.S., & PETINA, A.A. (1974) [Selenium and tellurium levels in environmental media.] In: [Hygiene and occupational diseases,] Moscow, Nauka Publishing House, Vol 21, pp. 69-71 (in Russian).

SENF, H.W. (1941) [A case of poisoning by hydrogen selenide]. Dtsch. Med. Wochenschr., 67: 1094-1096 (in German).

SEWARD, C.R., VAUGHAN, G., & HOVE, E.L. (1966) Effect of selenium on incisor depigmentation and carbon tetrachloride poisoning in vitamin E-deficient rats. Proc. Soc. Exp. Biol. Med., 121: 850-852.

SHACKLETTE, H.T., BOERNGEN, J.G., & KEITH, J.R. (1974) Selenium, fluorine, and arsenic in surficial materials of the conterminous United States, 14 pp (US Geological Survey Circular, 692).

SHAMBERGER, R.J. (1970) Relationship of selenium to cancer. I. Inhibitory effect of selenium on carcinogenesis. J. Natl Cancer Inst., 44: 931-936.

SHAMBERGER, R.J. (1971) Is selenium a teratogen? Lancet, 2: 1316.

SHAMBERGER, R.J. (1978) Antioxidants and cancer. VIII. Cadmium-selenium levels in kidneys. In: Kirchgessner, M., ed. Trace element metabolism in man and animals. III, Freising-Weihenstephan, West Germany, Arbeitskreis für Tierernährungsforschung Weihenstephan, pp. 391-392.

SHAMBERGER, R.J. & FROST, D.V. (1969) Possible protective effect of selenium against human cancer. Can. Med. Assoc. J., 100: 682.

SHAMBERGER, R.J. & RUDOLPH, G. (1966) Protection against cocarcinogenesis by antioxidants. Experientia (Basel), 22: 116.

SHAMBERGER, R.J. & WILLIS, C.E. (1971) Selenium distribution and human cancer mortality. Crit. Rev. clin. Lab. Sci., 2: 211-221.

SHAMBERGER, R.J., BAUGHMAN, F.F., KALCHERT, S.L., WILLIS, C.E., & HOFFMAN, G.C. (1973a) Carcinogen-induced chromosomal breakage decreased by antioxidants. Proc. Natl Acad. Sci. (USA), 70: 1461-1463.

SHAMBERGER, R.J., RUKOVENA, E., LONGFIELD, A.K., TYTKO, S.A., DEODHAR, S., & WILLIS, C.E. (1973b) Antioxidants and cancer. I. Selenium in the blood of normals and cancer patients. J. Natl Cancer Inst., 50(4): 863-870.

SHAMBERGER, R.J., TYTKO, S.A., & WILLIS, C.E. (1975) Selenium and heart disease. In: Hemphill, D.D., ed. Trace substances in environmental health. IX, Columbia, Missouri, University of Missouri Press, pp. 15-22.

SHAMBERGER, R.J., TYTKO, S.A., & WILLIS, C.E. (1976) Antioxidants and cancer. VI. Selenium and age-adjusted human cancer mortality. Arch. environ. Health, 31: 231-235.

SHAMBERGER, R.J., CORLETT, C.L., BEAMAN, K.D., & KASTEN, B.L. (1979) Antioxidants reduce the mutagenic effect of malonaldehyde and β-propiolactone. Partix, antioxidants, and cancer. Mutat. Res., 66: 349-355.

SHAMBERGER, R.J., WILLIS, C.E., & MCCORMACK, L.J. (1979) Selenium and heart disease. III. Blood selenium and heart mortality in 19 states. In: Hemphill, D.D., ed. Trace substances in environmental health, Columbia, Missouri, University of Missouri, pp. 59-63.

SHAPIRO, J.R. (1972) Selenium and carcinogenesis: a review. Ann. NY Acad. Sci., 192: 215-219.

SHEARER, T.R. (1973) Lack of effect of selenium on glycolytic and citric acid cycle intermediates in rat kidney and liver. Proc. Soc. Exp. Biol. Med., 144: 688-691.

SHEARER, T.R. (1975) Developmental and postdevelopmental uptake of dietary organic and inorganic selenium into the molar teeth of rats. J. Nutr., 105: 338-347.

SHEARER, T.R. & HADJIMARKOS, D.M. (1975) Geographic distribution of selenium in human milk. Arch. environ. Health, 30: 230-233.

SHEARER, T.R. & RIDLINGTON, J.W. (1976) Fluoride-selenium interaction in the hard and soft tissues of the rat. J. Nutr., 106: 451-456.

SHEARER, T.R., MCCORMACK, D.W., DESART, D.J., BRITTON, J.L., & LOPEZ, M.T. (1980) Histological evaluation of selenium induced cataracts. Exp. Eye Res., 31: 327-333.

SHENDRIKAR, A.D. (1974) Critical evaluation of analytical methods for the determination of selenium in air, water, and biological materials. Sci. total Environ., 3: 155-168.

SHENDRIKAR, A.D. & FAUDEL, G.B. (1978) Distribution of trace metals during oil shale retorting. Environ. Sci. Technol., 12: 332-334.

SHRIFT, A. (1964) A selenium cycle in nature? Nature (Lond.), 201: 1304-1305.

SHRIFT, A. (1973) Selenium compounds in nature and medicine. Metabolism of selenium by plants and microorganisms. In: Klayman, D.L. & Gunther, W.H.H., ed. Organic selenium compounds: their chemistry and biology, New York, John Wiley and Sons, Inc, pp. 763-814.

SHRIFT, A. & VIRUPAKSHA, T.K. (1965) Seleno-amino acids in selenium-accumulating plants. Biochim. Biophys. Acta, 100: 67-75.

SHUM, G.T.C., FREEMAN, H.C., & UTHE, J.F. (1977) Flameless atomic absorption spectrophotometry of selenium in fish and food products. J. Assoc. Off. Anal. Chem., 60: 1010-1014.

SHUMAEV, V.D., MAKUSHINSKAJA, N.D., & CHIGRINA, T.A. (1976) [Hygiene aspects of industrial waste water pollution in some non-ferrous metal enterprises in Kazahstan by chemicals regulated for the point of sanitary toxicological risk characteristics.] Zolvavoohram Kazah, 4: 36-38 (in Russian).

SIMPSON, B.H. (1972a) An epidemiological study of carcinoma of the small intestine in New Zealand sheep. N.Z. vet. J., 20: 91-97.

SIMPSON, B.H. (1972b) The possible relationship of selenium and superphosphate to the frequencey of occurrence of intestinal carcinomas in sheep. N.Z. vet. J., 20: 224.

SINDEEVA, N.D. (1959) [Minerology, occurrence and main characteristics of the geochemistry of selenium and tellurium,] Moscow, Publishing House of the Academy of Sciences of the USSR, 257 pp (in Russian).

SIRIANNI, S.R. & HUANG, C.C. (1983) Induction of sister chromatid exchange by various selenium compounds in Chinese hamster cells in the presence and absence of S9 mixture. Cancer Lett., 18: 109-116.

SIVERTSEN, T., KARLSEN, J.T., & FROSLIE, A. (1977) The relationship of erythrocyte glutathione peroxidase to blood selenium in swine. Acta vet. Scand., 18: 494-500.

SKORNJAKOVA, L.V., BURCHANOV, A.I., & SALECHOV, M.I. (1969) [On the acute manifestations of selenium poisoning.] Gig. Tr. Prof. Zabol., 11: 45-46 (in Russian).

SMITH, C.R., Jr, WEISLEDER, D., MILLER, R.W., PALMER, I.S., & OLSON, O.E. (1980) Linustatin and neolinustatin: cyanogenic glycosides of linseed meal that protect animals against selenium toxicity. J. org. Chem., 45: 507-510.

SMITH, M.I. & STOHLMAN, E.F. (1940) Further observations on the influence of dietary protein on the toxicity of selenium. J. Pharmacol. exp. Ther., 70: 270-278.

SMITH, M.I. & WESTFALL, B.B. (1937) Further field studies on the selenium problem in relation to public health. US Public Health Rep., 52: 1375-1384.

SMITH, M.I., FRANKE, K.W., & WESTFALL, B.B. (1936) The selenium problem in relation to public health. A preliminary survey to determine the possibility of selenium intoxication in the rural population living on seleniferous soil. US Public Health Rep., 51: 1496-1505.

SMITH, M.I., WESTFALL, B.B., & STOHLMAN, E.F., Jr (1937) The elmination of selenium and its distribution in the tissues. US Public Health Rep., 52: 1171-1177.

SMITH, M.I., WESTFALL, B.B., & STOHLMAN, E.F., Jr (1938) Studies on the fate of selenium in the organism. US Public Health Rep., 53: 1199-1216.

SOKOLOFF, L. (1985) Endemic forms of osteoarthritis. Clin. rheum. Dis., 11: 187-202.

SPALLHOLZ, J.E., MARTIN, J.L., GERLACH, M.L., & HEINZERLING, R.H. (1973) Immunologic responses of mice fed diets supplemented with selenite selenium. Proc. Soc. Exp. Biol. Med., 143: 685-689.

SPALLHOLZ, J.E., MARTIN, J.L., GERLACH, M.L., & HEINZERLING, R.H. (1975) Injectable selenium: effect on the primary immune response of mice. Proc. Soc. Exp. Biol. Med., 148: 37-40.

SPALLHOLZ, J.E., COLLINS, G.E., & SCHWARZ, K. (1978) A single test-tube method for the fluorometric microdetermination of selenium. Bioinorg. Chem., 9: 453-459.

SPRINKER, L.H., HARR, J.R., NEWBERNE, P.M., WHANGER, P.D., & WESWIG, P.H. (1971) Selenium deficiency lesions in rats fed vitamin E supplemented rations. Nutr. Rep. Int., 4: 335-340.

STADTMAN, T.C. (1977) Biological function of selenium. Nutr. Rev., 35: 161-166.

STEAD, R.J., HINKS, L.J., HODSON, M.E., REDINGTON, A.N., CLAYTON, B.E., & BATTEN, J.C. (1985) Selenium deficiency and possible increased risk of carcinoma in adults with cystic fibrosis. Lancet(October): 862-863.

STEWART, R.D.H., GRIFFITHS, N.M., THOMSON, C.D., & ROBINSON, M.F. (1978) Quantitative selenium metabolism in normal New Zealand women. Br. J. Nutr., 40: 45-54.

STOEWSAND, G.S., GUTENMANN, W.H., & LISK, D.J. (1978) Wheat grown on fly ash: high selenium uptake and response when fed to Japanese quail. J. agric. food Chem., 26: 757-759.

SU, Y. & YU, W. (1983) Nutritional bio-geochemical etiology of Keshan disease. Chin. med. J., 96: 594-596.

SU, Y., CUI, S., GU, B., ZENG, X., & YU, W. (1982) Experimental study of the effects of corn and vegetables from Keshan disease endemic districts on the growth and myocardium in rats. Acta nutr. Sinica, 4: 261-269.

SUBCOMMITTEE ON SELENIUM - COMMITTEE ON ANIMAL NUTRITION (1983) Selenium in nutrition, revised ed., Washington DC, Board on Agriculture, National Research Council, National Academy of Sciences, 174 pp.

SUCHKOV, B.P. (1971) [Selenium content in major nutrients consumed by the population of the Ukrainian SSR.] Vopr. Pitan., 30(b): 75-77 (in Russian).

SUCHKOV, B.P. & KACAP, I.M. (1971) [Dental caries in the population of the Chernovitsk region in relation with the impact of microelements.] Gig. i Sanit., 3: 91-93 (in Russian).

SUCHKOV, B.P. & ZHIVECKIJ, A.V. (1973) [Selenium levels in the blood of healthy people and people with neoplasms in a Chernovici region republican interdepartmental collective.] In: [Microelements in medicine,] Kiev, Zdarovie, Zdarovie Publishing House, Vol. 4, pp. 69-73 (in Russian).

SUCHKOV, B.P., KACAP, I.M., & GULGASENKO, A.I. (1973) [Affection of the population of the Chernovitsi region with caries in association with selenium content in the teeth]. Stomatologia, 52: 21-23 (in Russian).

SUCHKOV, B.P., SHEVCHUK, I.À., & MARBAR, A.I. (1977) [Histopathomorphological changes in the organs and tissues of laboratory animals on a synthetic diet with a low content of selenium and vitamin E.] Vopr. Pitan., 2: 33-41 (in Russian).

SUCHKOV, B.P., SHTUTMAN, C.M., & HALMURADOV, A.G. (1978)
⌊The biochemical role of selenium in animal organisms.⌋
Ukrain. Biochem., 50: 659-671 (in Russian).

SUN, S., ZHAI, F., ZHOU, L., & YANG, G. (1985) The
bioavailability of soil selenium in Keshan disease and high
selenium areas. Chinese J. end. Dis., 4: 21-28.

SUNDE, R.A. (1984) Selenoproteins. J. Am. Oil Chem. Soc.,
61: 1891-1898.

SUNDSTROM, H., KORPELA, H., VIINIKKA, L., & KAUPPILA, A.
(1984a) Serum selenium and glutathione peroxidase, and plasma
lipid peroxides in uterine, ovarian or vulvar cancer, and
their responses to antioxidants in patients with ovarian
cancer. Cancer Lett., 24: 1-10.

SUNDSTROM, H., YRJANHEIKKI, E., & KAUPPILA, A. (1984b) Serum
selenium in patients with ovarian cancer during and after
therapy. Carcinogenesis, 5(6): 731-734.

SVERDLINA, N.T & MASLENNIKOVA, V.S. (1961) ⌊Hygienic
evaluation of working conditions in production of selenium
photocells.⌋ Leningrad, Mater. manch sessin Leningrad NII gig.
truda prof. Zab, pp. 73-75 (in Russian).

SWEINS, A. (1983) Protective effect of selenium against
arsenic-induced chromosomal damage in cultured human lymphocytes. Hereditas, 98: 249-252.

SYMANSKI, H. (1950) ⌊A case of hygrogen selenide poisoning.⌋
Dtsch. Med. Wochenschr., 75: 1730 (in German).

SZYDLOWSKI, F.J. & DUNMIRE, D.L. (1979) Semi-automatic
digestion and automatic analysis for selenium in animal feeds.
Anal. Chim Acta, 105: 445-449.

TAN, J., ZHENG, D., HOU, S., ZHU, W., LI, R., ZHU, Z., & WANG,
W. (in press) Selenium ecological chemico-geography and
endemic Keshan disease and Kashin-Beck disease in China. In:
Proceedings of the Third International Symposium on Selenium
in Biology and Medicine.

TANK, G. & STORVICK, C.A. (1960) Effect of naturally
occurring selenium and vanadium on dental caries. J. dent.
Res., 39: 473-488.

TAYLOR, F.B. (1963) Significance of trace elements in public finished water supplies. J. Am. Water Works Assoc., 55: 619-623.

TEEL, R.W. (1984) A comparison of the effect of selenium on the mutagenicity and metabolism of benzo(α)pyrene in rat and hamster liver S9 activation systems. Cancer Lett., 24: 281-289.

TEEL, R.W. & KAIN, S.R. (1984) Selenium modified mutagenicity and metabolism of benzo(α)pyrene in an S9-dependent system. Mutat. Res., 127: 9-14.

THOMPSON, H.J. (1984) Selenium as an anticarcinogen. J. agric. food Chem., 32: 422-425.

THOMPSON, J.N. & SCOTT, M.L. (1969) Role of selenium in the nutrition of the chick. J. Nutr., 97: 335-342.

THOMPSON, J.N. & SCOTT, M.L. (1970) Impaired lipid and vitamin E absorption related to atrophy of the pancreas in selenium-deficient chicks. J. Nutr., 100: 797-809.

THOMPSON, J.N., ERDODY, P., & SMITH, D.C. (1975) Selenium content of food consumed by Canadians. J. Nutr., 105: 274-277.

THOMPSON, K.C. (1975) The atomic-fluorescence determination of antimony, arsenic, selenium and tellurium by using the hydride generation technique. Analyst, 100: 307-310.

THOMPSON, R.H., MCMURRAY, C.H., & BLANCHFLOWER, W.J. (1976) The levels of selenium and glutathione peroxidase activity in blood of sheep, cows and pigs. Res. vet. Sci., 20: 229-231.

THOMSON, C.D. (1974) Recovery of large doses of selenium given as sodium selenite with or without vitamin E. N.Z. med. J., 80: 163-168.

THOMSON, C.D. (1985) Selenium dependent and non-selenium dependent glutathione peroxidase in human tissues of New Zealand residents. Biochem. Int., 10: 673-679.

THOMSON, C.D. & ROBINSON, M.F. (1980) Selenium in human health with emphasis on those aspects peculiar to New Zealand. Am. J. clin. Nutr., 33: 303-323.

THOMSON, C.D. & ROBINSON, M.F. (1986) Urinary and faecal excretions and absorption of a large supplement of selenium as selenite or as selenate. Am. J. clin. Nutr. (submitted for publication).

THOMSON, C.D. & STEWART, R.D.H. (1973) Metabolic studies of ^{75}Se selenomethionine and ^{75}Se selenite in the rat. Br. J. Nutr., 30: 139-147.

THOMSON, C.D. & STEWART, R.D.H. (1974) The metabolism of ^{75}Se selenite in young women. Br. J. Nutr., 32: 47-57.

THOMSON, C.D., ROBINSON, B.A., STEWART, R.D.H., & ROBINSON, M.F. (1975a) Metabolic studies of ^{75}Se selenocystine and ^{75}Se selenomethionine in the rat. Br. J. Nutr., 34: 501-509.

THOMSON, C.D., STEWART, R.D.H., & ROBINSON, M.F. (1975b) Metabolic studies in rats of ^{75}Se selenomethionine and of ^{75}Se incorporated in vivo into rabbit kidney. Br. J. Nutr., 33: 45-54.

THOMSON, C.D., REA, H.M., ROBINSON, M.F., & CHAPMAN, O.W. (1977a) Low blood selenium concentrations and glutathione peroxidase activities in elderly people. Proc. Univ. Otago Med. Sch., 55: 18-19.

THOMSON, C.D., REA, H.M., DOESBURG, V.M., & ROBINSON, M.F. (1977b) Selenium concentrations and glutathione peroxidase activities in whole blood of New Zealand residents. Br. J. Nutr., 37: 457-460.

THOMSON, C.D., BURTON, C.E., & ROBINSON, M.F. (1978) On supplementing the selenium intake of New Zealanders. I. Short experiments with large doses of selenite or selenomethionine. Br. J. Nutr., 39: 579-587.

THOMSON, C.D., ROBINSON, M.F., CAMPBELL, D.R., & REA, H.M. (1982) Effect of prolonged supplementation with daily supplements of selenomethionine and sodium selenite on glutathione peroxidase activity in blood of New Zealand residents. Am. J. clin. Nutr., 36: 24-31.

THOMSON, C.D., ONG, L.K., & ROBINSON, M.F. (1985) Effect of supplementation with high-selenium wheat bread on selenium, glutathione peroxidase and related enzymes in blood components of New Zealand residents. Am. J. clin. Nutr., 41: 1015-1022.

THORN, J., ROBERTSON, J., & BUSS, D.H. (1978) Trace nutrients. Selenium in British food. Br. J. Nutr., 39: 391-396.

TINSLEY, I.J., HARR, J.R., BONE, J.F., WESWIG, P.H., & YAMAMOTO, R.S. (1967) Selenium toxicity in rats. I. Growth and longevity. In: Muth, O.H., Oldfield, J.E., & Weswig, P.H., ed. Selenium in biomedicine, Westport, Connecticut, The AVI Publishing Co., Inc, pp. 141-152.

TSEN, C.C. & COLLIER, H.B. (1959) Selenite as a relatively weak inhibitor of some sulphydryl enzymes. Nature (Lond.), 183: 1327-1328.

TSEN, C.C. & TAPPEL, A.L. (1958) Catalytic oxidation of glutathione and other sulfhydryl compounds by selenite. J. biol. Chem., 233(5): 1230-1232.

TSONGAS, T.A. & FERGUSON, S.W. (1977) Human health effects of selenium in a rural Colorado drinking-water supply. In: Hemphill, D.D., ed. Trace substances in envrionmental health. XI, Columbia, Missouri, University of Missouri Press, pp. 30-35.

ULLREY, D.E., BRADY, P.S., WHETTER, P.A., KU, P.K., & MAGEE, W.T. (1977) Selenium supplementation of diets for sheep and beef cattle. J. anal. Sci., 45: 559-565.

UNDERWOOD, E.J. (1977) Trace elements in human and animal nutrition, 4th ed., New York, Academic Press, 545 pp.

US DEPARTMENT OF HEALTH AND HUMAN SERVICES, FOOD AND DRUG ADMINISTRATION (1984) 21 CFR Pt.573.920 Selenium. Fed. Reg., 49: 627-628.

US FDA (1974) Final environmental impact statement rule making on selenium in animal feeds, Washington DC, US Food and Drug Administration, 131 pp.

US FDA (1975) Bureau of Foods Compliance Program Evaluation Report. Total diet studies, Fiscal Year 1974, Washington DC, US Food and Drug Administration, 32 pp.

US NAS/NRC (1971) Selenium in nutrition, Washington DC, National Academy of Science, National Research Council, Agricultural Board, Committee on Animal Nutrition, Subcommittee on Selenium, 79 pp.

US NAS/NRC (1976) Selenium, Washington DC, National Academy of Science, National Research Council, Assembly of Life Sciences, Medical and Biological Effects of Environmental Pollutants, 203 pp.

US NAS/NRC (1979) <u>Zinc</u>, Baltimore, Maryland, National Academy of Science, National Research Council, Assembly of Life Sciences, Committee on Medical and Biologic Effects of Environmental Pollutants, 471 pp.

US NAS/NRC (1980) <u>Recommended dietary allowances</u>, Washington DC, National Academy of Science, National Research Council, Food and Nutrition Board, Committee on Dietary Allowances, 185 pp.

VAN DER LINDEN, R., DE CORTE, F., & HOSTE, J. (1974) Activation analysis of biological material with ruthenium as a multi-element comparator. <u>Anal. Chim. Acta</u>, <u>71</u>: 263-275.

VAN KAMPEN, K.R. & JAMES, L.F. (1978) Manifestations of intoxication by selenium-accumulating plants. In: Keeler, R.F., Van Kampen, K.R., & James, L.F., ed. <u>Effects of poisonous plants on livestock</u>, New York, Academic Press, pp. 135-138.

VAN RIJ, A.M., ROBINSON, M.F., GODFREY, P.J., THOMSON, C.D., & RHEA, H.M. (1978) Selenium blood levels in cancer and other diseases in surgery. In: Hemphill, D.D., ed. <u>Trace substances in environmental health. XII</u>, Columbia, Missouri, University of Missouri Press, pp. 157-163.

VAN RIJ, A.M., THOMSON, C.D., MCKENZIE, J.M., & ROBINSON, M.F. (1979) Selenium deficiency in total parenteral nutrition. <u>Am. J. clin. Nutr.</u>, <u>32</u>: 2076-2085.

VAN VLEET, J.F. (1976) Induction of lesions of selenium-vitamin E deficiency in pigs fed silver. <u>Am. J. vet. Res.</u>, <u>37</u>: 1415-1420.

VAN VLEET, J.F., MEYER, K.B., & OLANDER, H.J. (1974) Acute selenium toxicosis induced in baby pigs by parenteral administration of selenium-vitamin E preparations. <u>J. Am. Vet. Med. Assoc.</u>, <u>165</u>: 543-547.

VARO, P. & KOIVISTOINEN, P. (1980) Mineral element composition of Finnish foods: N, K, Ca, Mg, P, S, Fe, Cu, Mn, Zn, Mo, Co, Ni, Cr, F, Se, Si, Rb, Al, B, Br, Hg, As, Cd, Pb, and ash. XII. General discussion and nutritional evaluation. <u>Acta agric. Scand. Suppl.</u>, <u>22</u>: 165.

VARO, P. & KOIVISTOINEN, P. (1981) Annual variations in the average selenium intake in Finland: cereal products and milk as sources of selenium in 1979/80. <u>Int. J. Vit. Nutr. Res.</u>, <u>51</u>: 79-84.

VERNIE, L.N. (1984) Selenium in carcinogenesis. Biochim. Biophys. Acta, 738(4): 203-217.

VERNIE, L.N., BONT, W.S., GINJAAR, H.B., & EMMELOT, P. (1975) Elongation factor 2 as the target of the reaction product between sodium selenite and glutathione (GSSeSG) in the inhibiting of amino acid incorporation in vitro. Biochim. Biophys. Acta, 414: 283-292.

VERNIE, L.N., GINJARR, H.B., WILDERS, I.T., & BONT, W.S. (1978) Amino acid incorporation in a cell-free system derived from rat liver studied with the aid of selenodiglutathione. Biochim. Biophys. Acta, 518(3): 507-517.

VIRTAMO, J., VALKEILA, E., ALFTHAN, G., PUNSAR, S., HUTTUNEN, J.K., & KARVONEN, M. (1985) Serum selenium and the risk of coronary heart disease and stroke. Am. J. Epidemiol., 122(2): 276-282.

VOBECKY, M., PAVLIK, L., & BENES, J. (1977) Non-destructive neutron activation assay of submicrogram quantities of selenium. Radiochem. Radioanal. Lett., 29(4): 159-164.

VOBECKY, M., DEDINA, J., PAVLIK, L., & VALASEK, J. (1979) Gamma-ray interferences in the determination of selenium by the Inaa method. Radiochem. Radioanal. Lett., 38(3): 197-204.

VOLGAREV, M.N. & TSCHERKES, L.A. (1967) Further studies in tissue changes associated with sodium selenate. In: Muth, O.H., Oldfield, J.E., & Weswig, P.H., ed. Selenium in biomedicine, Westport, Connecticut, The AVI Publishing Co., Inc, pp. 179-184.

WAGNER, P.A., HOEKSTRA, W.G., & GANTHER, H.E. (1975) Alleviation of silver toxicity by selenite in the rat in relation to tissue glutathione peroxidase. Proc. Soc. Exp. Biol. Med., 148: 1106-1110.

WAHLSTROM, R.C. & OLSON, O.E. (1959) The effect of selenium on reproduction in swine. J. anim. Sci., 18: 141-145.

WALKER, G.W.R. & BRADLEY, A.M. (1969) Interacting effects of sodium monohydrogenarsenate and selenocystine on crossing over in Drosophila melanogaster. Can. J. Genet. Cytol., 11: 677-688.

WALKER, G.W.R. & TING, K.P. (1967) Effect of selenium on recombination in barley. Can. J. Genet. Cytol., 9: 314-320.

WANG, Z., LI, C., & LI, L. (1985) An epidemiological investigation on the selenium content of water, cereals and hair of children in Heilongjiang province. Chinese J. end. Dis., 4: 330-333.

WARRINGTON, P.B. (1979) Selenium. Eng. Min. J., 180: 149-150.

WASLIEN, C.I. (1976) Human intake of trace elements. In: Prasad, A.S., ed. Trace elements in human health and disease. II. Essential and toxic elements, New York, Academic Press, pp. 347-370.

WATKINSON, J.H. (1967) Analytical methods for selenium in biological material. In: Muth, O.H., Oldfield, J.E., & Weswig, P.H., ed. Selenium in biomedicine, Westport, Connecticut, The AVI publishing Co., Inc, pp. 97-117.

WATKINSON, J.H. (1974) The selenium status of New Zealanders. N.Z. med. J., 80: 202-205.

WATKINSON, J.H. (1979) Semi-automated fluorimetric determination of nanogram quantities of selenium in biological materials. Anal. Chim. Acta, 105: 319-325.

WATKINSON, J.H. (1981) Changes of blood selenium in New Zealand adults with time and importation of Australian wheat. Am. J. clin. Nutr., 34: 936-942.

WATKINSON, J.H. & BROWN, M.W. (1979) New phase-separating device and other improvements in the semi-automated fluorimetric determination of selenium. Anal. Chim. Acta, 105: 451-454.

WEDDERBURN, J.F. (1972) Selenium and cancer. N.Z. vet. J., 20: 56-57.

WEISSMAN, S.H., CUDDIHY, R.G., & BURKSTALLER, M.A. (1979) Distribution and retention of inhaled selenious acid and selenium metal aerosols in Beagle dogs. In: Hemphill, D.D., ed. Trace substances in environmental health. XIII, Columbia, Missouri, University of Missouri Press, pp. 477-482.

WEISSMAN, S.H., CUDDIHY, R.G., & MEDINSKY, M.A. (1983) Absorption, distribution, and retention of inhaled selenious acid and selenium metal aerosols in Beagle dogs. Toxicol. appl. Pharmacol., 67: 331-337.

WELSH, S.O. (1979) The protective effect of vitamin E and N,N'-diphenyl-p-phenylenediamine (DPPD) against methyl mercury toxicity in the rat. J. Nutr., 109: 1673-1681.

WELSH, S.O. & SOARES, J.H., Jr (1976) The protective effect of vitamin E and selenium against methyl mercury toxicity in the Japanese quail. Nutr. Rep. Int., 13: 43-51.

WELSH, S.O., HOLDEN, J.W., WOLF, W.R., & LEVANDER, O.A. (1981) Selenium intake of Maryland residents consuming self-selected diets. J. Am. Diet. Assoc., 79: 277-285.

WESTERMANN, D.T. & ROBBINS, C.W. (1974) Effect of SO₄-S fertilization on Se concentration of alfalfa (Medicago sativa L). Agron. J., 66: 207-208.

WESTERMARCK, T. (1977) Selenium content of tissues in Finnish infants and adults with various diseases, and studies on the effects of selenium supplementation in neuronal ceroid lipofuscinosis patients. Acta pharmacol. toxicol., 41: 121-128.

WESTERMARCK, T., RAUNU, P., KIRJARINTA, M., & LAPPALAINEN, L. (1977) Selenium content of whole blood and serum in adults and children of different ages from different parts of Finland. Acta pharmacol. toxicol., 40: 465-475.

WESWIG, P.H., TINSLEY, I.J., HARR, J.R., BONE, J.F., YAMAMOTO, R.S., & FALK, H. (1966) Bioassay of selenium compounds for carcinogensis in rats. Final report, Corvallis, Oregon State University, Departments of Agricultral Chemistry and Veterinary Medicine, 131 pp.

WHANGER, P.D. (1976) Selenium versus metal toxicity in animals. In: Proceedings of the Symposium on Selenium-Tellurium in the Environment, Pittsburgh, Pennsylvania, Industrial Health Foundation, pp. 234-252.

WHANGER, P.D., MUTH, O.H., OLDFIELD, J.E., & WESWIG, P.H. (1969) Influence of sulfur on incidence of white muscle disease in lambs. J. Nutr., 97: 553-562.

WHANGER, P.D., PEDERSEN, N.D., & WESWIG, P.H. (1973) Selenium proteins in ovine tissues. II. Spectral properties of a 10,000 molecular weight selenium protein. Biochem. Biophys. Res. Commun., 53: 1031-1035.

WHANGER, P.D., PEDERSON, N.D., HATFIELD, J., & WESWIG, P.H. (1976a) Absorption of selenite and selenomethionine from ligated digestive tract segments in rats. Proc. Soc. Exp. Biol. Med., 153: 295-297.

WHANGER, P.D., WESWIG, P.H., SCHMITZ, J.A., & OLDFIELD, J.E. (1976b) Effects of selenium, cadmium, mercury, tellurium, arsenic, silver, and cobalt on White Muscle Disease in lambs and effect of dietary forms of arsenic on its accumulation in tissues. Nutr. Rep. Int., 14: 63-72.

WHEATCROFT, M.G., ENGLISH, J.A., & SCHLACK, C.A. (1951) Effects of selenium on the incidence of dental caries in white rats. J. dent. Res., 30: 523-524.

WHETTER, P.A. & ULLREY, D.E. (1978) Improved fluorometric method for determining selenium. J. Assoc. Off. Anal. Chem., 61: 927-930.

WHITEACRE, M.E., COMBS, G.F., & PARKER, R.S. (1983) Peroxidative damage in nutritional pancreatic atrophy due to selenium-deficiency in the chick. Fed. Proc. Fed. Am. Soc. Exp. Biol., 42: 928.

WHITING, R.F., WEI, L., & STICH, H.F. (1980) Unscheduled DNA synthesis and chromosome aberrations induced by inorganic and organic selenium compounds in the presence of glutathione. Mutat. Res., 78: 159-169.

WHO (1984) Guidelines for drinking water quality. Volume 2: Health criteria and other supporting information, Geneva, World Health Organization, 335 pp.

WILKIE, J.B. & YOUNG, M. (1970) Improvement in the 2,3-diaminoaphthalene reagent for microfluorescent. Determination of selenium in biological materials. J. agric. food Chem., 18: 944-945.

WILLETT, W.C., MORRIS, J.S., PRESSEL, S., TAYLOR, J.O., POLK, B.F., STAMPFER, M.J., ROSNER, B., SCHNEIDER, K., & HAMES, C.G. (1983) Prediagnostic serum selenium and risk of cancer. Lancet(July): 130-134.

WILLIAMS, K.T. & BYERS, H.G. (1935) Occurrence of selenium in the Colorado River and some of its tributaries. Ind. Eng. Chem. Anal. Ed., 7: 431-432.

WILLIAMS, M.M.I. (1983) Selenium and glutathione peroxidase in mature human milk. Proc. Univ. Otago Med. Sch., 61: 20-21.

WILSON, H.M. (1962) Selenium oxide poisoning. N.C. med. J., 23: 73-75.

WILSON, P.S. & JUDSON, G.J. (1976) Glutathione peroxidase activity in bovine and ovine erythrocytes in relation to blood selenium concentration. Br. vet. J., 132: 428-434.

WITTING, L.A. & HORWITT, M.K. (1964) Effects of dietary selenium, methionine, fat level, and tocopherol on rat growth. J. Nutr., 84: 351-360.

WRIGHT, P.L. & BELL, M.C. (1966) Comparative metabolism of selenium and tellurium in sheep and swine. Am. J. Physiol., 211: 6-10.

WU, A.S.H., OLDFIELD, J.E., & WHANGER, P.D. (1971) Effect of selenium, chromium and vitamin E on spermatogensis. J. anal. Sci., 33: 273.

WU, A.S.H., OLDFIELD, J.E., WHANGER, P.D., & WESWIG, P.H. (1973) Effect of selenium, vitamin E, and antioxidants on testicular function in rats. Biol. Reprod., 8: 625-629.

WU, A.S.H., OLDFIELD, J.E., SHULL, L.R., & CHEEKE, P.R. (1979) Specific effect of selenium deficiency on rat sperm. Biol. Reprod., 20: 793-798.

XIA, X. & YU, S. (in press) Studies on the anti-carcinogenic mechanisms of selenium - The effects of glucose oxidation and related enzymes. Chinese J. Cancer.

YANG, G.Q., WANG, S., ZHOU, R., & SUN, S. (1983) Endemic selenium intoxication of humans in China. Am. J. clin. Nutr., 37: 872-881.

YANG, G., CHEN, J., WEN, Z., GE, K., ZHU, L., CHEN, X., & CHEN, X. (1984) The role of selenium in Keshan disease. In: Draper, H.H., ed. Advances in nutritional research, New York, Plenum Press, pp. 203-231.

YANG, G., ZHOU, R., YIN, S., PIAO, J., ZHU, L., LIN, S., & GU, L., MAI, R., XU, J., QIAG, F., LU, M., DENG, X., HUANG, J., LIN, & ZHOU, W. (1985) [Studies on the selenium requirement of Chinese people. I. Physiological requirement, minimum requirement, and minimum allowance.] J. Inst. Health, 14: 24-28 (in Chinese).

YANG, G.Q., ZHU, L.Z., LIU, S.J., GU, L.Z., QIAN, P.C., HUANG, J.H., & LU, M.D. (1986) Studies of human selenium requirements in China. In: Combs, G.F., Spallholz, J.E., Levander, O.A., & Oldfield, J.E., ed. Selenium in biology and medicine, 3rd Symposium, Westport, Connecticut, AVI Publishing Company.

YASUMOTO, K., IWAMI, K., YOSHIDA, M., & MITSUDA, H. (1976) Selenium content of foods and its average daily intake in Japan. Eiyo To Shokuryo, 29: 511-515.

YEH, Y. & JOHNSON, R.M. (1973) Vitamin E deficiency in the rat. IV. Alteration in mitochondrial membrane and its relation to respiratory decline. Arch. Biochem. Biophys., 159: 821-831.

YOUNG, S. & KEELER, R.F. (1962) Nutritional muscular dystrophy in lambs. The effect of muscular activity on the symmetrical distribution of lesions. Am. J. vet. Res., 23: 966-971.

YU, W.H. (1982) A study of nutritional and bio-geochemical factors in the occurrence and development of Keshan disease. Jpn Circ. J., 46: 1201-1207.

YU, S., CHU, Y., GONG, X., & HOU, C. (1985) Region variation of cancer mortality incidence and its relation to selenium levels in China. Ecol. trace Element Res., 7: 21-23.

YU, S., ZHU, Y., HON, C., & HUANG, C. (in press) Selective effects of selenium on the function and structure of mitochondria isolated from hepatoma and normal liver. In: Proceedings of the Third International Symposium of Selenium in Biology and Medicine.

ZABEL, N.L., HARLAND, J., GORMICAN, A.T., & GANTHER, H.E. (1978) Selenium content of commercial formula diets. Am. J. clin. Nutr., 31: 850-858.

ZHU, L. & LU, Z. et al. (1981) Effects in pigs fed the crops grown in Keshan disease affected province of China. In: Howell, J.M., Gawthone, J.M., & White, C.L., ed. Trace Element Metabolism in Man and Animals (TEMA 4).

ZOLLER, W.H. & REAMER, D.C. (1976) Selenium in the atmosphere. In: Proceedings of the Symposium on Selenium-Tellurium in the Environment, Pittsburgh, Pennsylvania, Industrial Health Foundation, pp. 54-66.

ZOLLER, W.H., GLADNEY, E.S., & DUCE, R.A. (1974) Atmospheric concentrations and sources of trace metals at the South Pole. Science, 183: 198-200.